高等院校土建学科双语教材（中英文对照）
◆ 工程管理专业 ◆
BASICS

工程项目规划

[德] 哈特穆特·克莱因　编著
赵　鹏　译

中国建筑工业出版社

著作权合同登记图字：01-2009-7708 号

图书在版编目（CIP）数据

工程项目规划/（德）克莱因编著；赵鹏译. —北京：中国建筑工业出版社，2011
高等院校土建学科双语教材（中英文对照）◆ 工程管理专业 ◆
ISBN 978-7-112-12254-7

Ⅰ.①工… Ⅱ.①克…②赵… Ⅲ.①建筑工程-项目管理 Ⅳ.①TU71

中国版本图书馆 CIP 数据核字（2010）第 145340 号

Basics: Project Planning/Hartmut Klein
Copyright © 2008 Birkhäuser Verlag AG (Verlag für Architektur), P. O. Box 133, 4010 Basel, Switzerland
Chinese Translation Copyright © 2011 China Architecture & Building Press
All rights reserved.
本书经 Birkhäuser Verlag AG 出版社授权我社翻译出版

责任编辑：孙　炼
责任设计：陈　旭
责任校对：赵　颖

高等院校土建学科双语教材（中英文对照）
◆ 工程管理专业 ◆
工程项目规划
［德］哈特穆特·克莱因　编著
赵　鹏　译
*
中国建筑工业出版社出版、发行（北京西郊百万庄）
各地新华书店、建筑书店经销
北京嘉泰利德公司制版
北京云浩印刷有限责任公司印刷
*
开本：880×1230 毫米　1/32　印张：4¼　字数：136 千字
2011 年 4 月第一版　　2011 年 4 月第一次印刷
定价：**16.00** 元
ISBN 978-7-112-12254-7
　　　　（20270）
版权所有　翻印必究
如有印装质量问题，可寄本社退换
（邮政编码 100037）

中文部分目录

\\ 序言　5

\\ 简介　79

\\ 项目参与者　80
　　\\ 业主　80
　　\\ 建筑师　81
　　\\ 项目管理员　82
　　\\ 专业设计人员　82
　　\\ 专家组　84
　　\\ 相关部门和权力机构　85
　　\\ 承包人　86
　　\\ 合同框架　86
　　\\ 团队建设　88

\\ 概况和目标　91
　　\\ 项目成本　91
　　\\ 项目期限　94
　　\\ 项目质量　97

\\ 规划进程　100
　　\\ 项目决策　100
　　\\ 概念阶段　102
　　\\ 设计阶段　104
　　\\ 获取许可阶段　108
　　\\ 工作计划和完工质量　110
　　\\ 投标　113
　　　　\\ 详细标书（质量清单）　118
　　　　\\ 可操作标书（投标纲要）　120
　　　　\\ 标书框架　120
　　\\ 决标程序　123
　　\\ 施工　127
　　\\ 保修期　135

\\ 结语　135

\\ 附录　136
　　\\ 图片提供者　136
　　\\ 作者　136

TABLE OF CONTENTS

\\Foreword _7

\\Introduction _8

\\Project participants _11
    \\The client _11
    \\The architect _12
    \\The project controller _13
    \\Specialist planners _14
    \\Experts _15
    \\Departments and authorities _17
    \\Contractors _17
    \\Contract structures _18
    \\Team building _20

\\General conditions and aims _23
    \\Costs _23
    \\Deadlines _27
    \\Quality _30

\\Planning process _33
    \\Deciding on a project _33
    \\Concept phase _35
    \\Design phase _38
    \\Gaining permission _42
    \\Working plans and quality of finish _45
    \\Tendering _49
        \\Detailed tender (bill of quantities) _54
        \\Functional tender specification (tender program) _56
        \\Structuring a tender _56
    \\The award procedure _60
    \\Construction _65
    \\The warranty period _73

\\In conclusion _75

\\Appendix _77
    \\Picture credits _77
    \\The author _77

序言

　　建筑师，以及其他规划设计人士的工作，是一个综合性的职业，是一个受多方面影响和关联的统一多样性的职业。这样的工作内容以及那些建筑师与之交往和交流中的人，从创造性的设计阶段到建筑工地的管理阶段，从创建一个工作规划团队到缜密地阐明建筑各项合同，均存在着调整变化和逐步完善。这一系列工作使建筑师这一职业是如此的令人激动和多姿多彩。

　　对于一个学生或新手来说，尽管日后的工作经历可以提供一个了解规划实践和工地现场的机会，但通常他们对一个建筑师正规的工作日状况仍缺乏足够的理解。本书可以满足处于这一知识层面的学生和新手们的需求，并且以易于理解的介绍和说明方式阐述了建筑师的工作任务。**本书的目的之一是探索一个项目从调查基础期和第一个灵感迸发到最终完工并移交给业主的全部过程。所有的规划和施工阶段都在行文中有所展现，同时，也为每个工作步骤提供了最重要的背景因素**。这意味着在表现力和精准细节方面，设计质量是可以转化为建筑质量的。另外，建筑师也能够调整一些规划参数，例如会议成本估算和时间期限。

　　当然，本书并不能代替职业经验，能够提供的只是最初的、一般性的概论，得出这个概论是总结了一个工程项目进行过程中诸多意外涌现的事件。然而，本书又确实可以提供实际有效的资料，建立核心工作和连接环节的概念，以便帮助学生和新手们在日后的职业生涯中形成一个良好的理解力，帮助他们在未来的办公室和工程项目中找到自己的位置。

编者：贝尔特·比勒费尔德

FOREWORD

The work of architects and other planners is a complex professional field, subject to a whole variety of influences and approaches. These, and the people with whom the architect interacts and communicates, change and develop, from the creative design stage to supervising work on the building site, from setting up a working planning team to carefully formulating building contracts. This range of work makes the architect's profession an exciting and very varied one.

A student or novice in the profession will usually have little insight into an architect's normal working day, though work experience can provide an introduction to planning practices or building sites. *Basics Project Planning* meets students and novices at this level of knowledge and explains architects' project work, structured with readily comprehensible introductions and explanations. One of the book's primary aims is to explore the overall approach to a project from investigating basics and the first idea to completion and handing over to the client. All the planning and working phases are presented in context and with the most important background elements for the individual working steps. This means that designed quality can be translated into built quality in terms of expression and precise detail, and the architect can control parameters such as meeting cost estimates and deadlines.

Of course *Basics Project Planning* is no substitute for professional experience, and can provide only a first, general survey, given the number of matters that crop up while a project is being planned. But it does provide practical information and structures the key jobs and connections so that students and novices can build up a good understanding of their later field of work, and find their way into given office and project structures.

Bert Bielefeld
Editor

INTRODUCTION

A project starts with the intention to translate a three-dimensional idea, a need for space or a property investment, into built reality. This "project" suggests both the desire for a convincing and high-quality concept and also the intention to stay with the concept until it is realized and completed. Project planning aims to bring an intention once expressed to its conclusion, and to turn the idea into built reality.

<small>Client/ architect</small>

Every project has to be initiated by a client; this applies to building, too. The client commissions someone – an architect – to draw up a design for preparing, planning, supervising, and executing a building project. The core task for the classical architect is project planning from investigating basics via the design to planning the work, tendering, site management, and building completion. For both client and architect, building costs, keeping to deadlines, and the quality of the completed work are highly relevant.

<small>Idea and realization</small>

The first impetus for planning a project can come to an architect in a variety of ways. In many cases a building client or investor will go directly to an architect he or she knows, and explain ideas and requirements for a project in a more or less concrete form. Building projects are often awarded via competitions or specialist reports, with several architects' designs competing to be chosen by the client on the basis of a previously formulated tender.

But the reverse is also conceivable: the architect approaches potential clients and puts him-/herself forward for a possible commission within an acquisition process. This involves careful research in order to find suitable clients needing buildings, or likely to need them in the near future.

<small>Planning steps/ decision levels</small>

In order to progress from the project idea to the completion and use of the real building, the project has to be planned and worked out step by step, in increasingly complex detail. The original abstract idea is gradually fully formulated, concretized and implemented in phases.

The project takes shape, is put on paper, the first sketches are drawn up. The number of people involved increases, sketches become scale plans, plans become the basis for applications. After permissions have been issued by the authorities, tenders have to be invited from building firms and tradespeople, and commissions awarded to contractors. The start of building is the first step towards actually realizing the project. The aim of putting up a building becomes reality once the various trades have been successfully coordinated on the building site.

Various steps have to be taken if a building project is to be planned and implemented sensibly and with foresight. These will differ according to the particular project and its structure and size. But the course of events is similar in each case, and can be generalized, even when responsibilities can be allotted in various ways. Thus in German-speaking countries the architect is the responsible key figure from the design phase to handing over the building, whereas in North America and many other European countries, responsibility is handed over to other partners after the design phase.

Decision levels

The client has to make decisions on various levels according to the different project phases.

1. Deciding on the project: In order to decide to embark on a project at all, various parameters (e.g. plot of land, function, financial and schedule framework) have to be examined for their fundamental acceptability. Decisions about carrying out planning are made by bringing in the necessary parties involved in the planning.
2. Deciding on the concept: If the architect's initial ideas (supported by parameters like functional connections, statements about volume and area, and rough costings) are already available, the client must decide whether he or she wants to have this first concept developed further to match his or her intentions.
3. Deciding to submit the building application: Once the concept including the above-mentioned parameters has been worked out further, the client has to decide whether to submit the existing design for permission from the building authorities, as only limited modifications can be made once this step has been taken.
4. Deciding on implementation qualities: Once planning permission has been obtained, the realization phase is prepared. In this context, the client has to decide between a number of possible qualities and material surfaces for the building work. This decision is usually based on material tests, samples, descriptions, and statements about cost development.
5. Deciding about awarding building contracts: Realization documents are prepared on the basis of previous decisions and contractors' implementation submissions collected. Now the client has to decide which contractor should be awarded the work. He or she is supported here by the architect's assessments and recommendations.

PROJECT PARTICIPANTS

The two most important parties to the project are undoubtedly the commissioning client and the planning architect. But there are a number of other "parties to the project" who have to be included in the planning process according to the size and ambition of the planned building. › Fig. 1

THE CLIENT

The client is the person or entity on whose authority a building is planned or erected. Legally the client can be a natural person or a juridical person under civil or public law.

Client/user Client and user may well be the same person, depending on the project at hand. In this case the architect has to seek agreement or clarification of the building project from only one person. But in public building projects in particular, and sometimes with private developers as well, the architect often has two persons as opposite numbers whose aims may well not be identical. There can also be further decision levels, such as external financial

Individuals placing commissions:		Client	
			Project controller
General planners:	Architect		
Specialist planners:	Structural engineer	Building technology engineer	Electrical engineer
	Surveyor	Interior designer	Landscape architect etc.
Report writers:	Soil expert	Fire prevention expert	Thermal insulation expert etc.
Local authorities:	Building supervision department	Fire brigade	etc.

Fig. 1:
Project participants

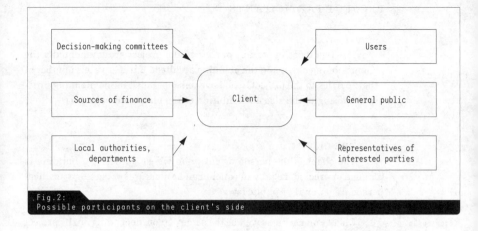

Fig.2:
Possible participants on the client's side

providers or supervisory committees on the client's side who can influence the planning and building process. The subsequent users depend on the project. They may be the teachers in a school, firefighters in a fire station, or doctors and nurses in a hospital. The client will generally involve them in planning at an early stage, but sometimes they can define requirements lying outside the client's scope. It is important to do justice to both sets of ideas and requirements if the planning is to be a success. › Fig. 2

THE ARCHITECT

In the construction field it is usually the architect or an expert planning company working in the building trade who will provide the required planning services.

Contacts/ representatives/ agents

The architect is the appropriate contact for all building questions. He or she advises the client on all matters appertaining to implementation, and works as his or her agent and representative with everyone involved in the building process: the authorities, other specialist planners or the firms and tradespeople carrying out the work.

Analysis, idea and solution

The architect examines the client's wishes critically in terms of their feasibility, gives advice and supplies ideas about the financial viability of the project, a realistic schedule estimate and possible design variants, thus working out possible approaches to solutions step by step with the client. Part of the architect's work is to develop convincing ideas and convey them successfully. A high level of successful teamwork is required when a large number of people are working on a project. Architects require a high

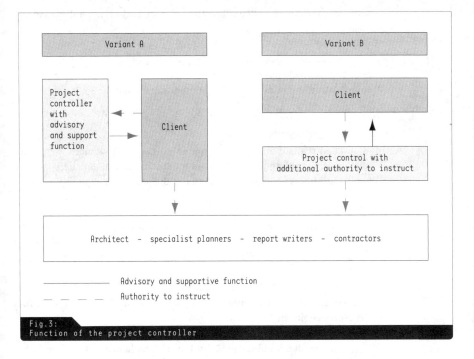

Fig.3:
Function of the project controller

degree of social competence and people skills to lead and guide all those involved. The various requirements within the individual planning phases are described in the planning process chapter.

THE PROJECT CONTROLLER

The number of people involved in a project as experts also increases with the size of the project. If the project size and time involved are so great as to exceed a client's capacities and expertise, it makes sense to involve a project controller.

The project controller takes over the technical, financial and legal client functions that can be delegated. In other words, he or she is also a client's advisor but does not usually have the authority to represent the client legally. ⟩ Fig. 3

Project management

The project controller's work does not usually relate to the architect's direct planning services (i.e. preliminary planning, design planning, etc.), but to management of the overall project, starting with financial analysis, providing resources and handling the contract, including facility management.

<small>Support for the architect</small>

For large projects in particular, the project controller can offer the architect valuable support in terms of project management, as well as in coordinating and controlling the personnel involved.

SPECIALIST PLANNERS

The architect will provide almost all planning services for a relatively small building project such as a detached house. But here as well, two other project partners are essential if the process is to be implemented correctly.

<small>Surveyor</small>

A chartered surveyor is generally commissioned to draw up an official site plan that will be required by the building authorities as part of the permission process, or will be needed later for the building survey.

<small>Structural engineer</small>

Then the statically relevant parts of the building – floor slab, walls, ceilings and roof – will be dimensioned by an appropriately qualified structural engineer.

For smaller building projects like a detached house the architect would usually also undertake the planning for electrical installations, heating, sanitation, and designing the outdoor areas; or work it out jointly with the firms and tradespeople commissioned to carry out the work.

When planning larger-scale building projects like public buildings (sports hall, town hall, fire station, etc.), large office complexes or prestigious company buildings, all the services will be provided by specialist engineers.

<small>Building services</small>

The engineer's contribution to providing services includes planning heating technology and sanitation, in other words water supply and sewage disposal. He or she will also take on planning for ventilation, cooling or air conditioning, and gas installations.

<small>Electrical engineer</small>

The electrical engineer does not just plan to supply the building with electricity and light, he or she is also jointly responsible for devising lightning protection, for fire and smoke alarms, and for signing escape routes.

Work that is usually carried out by the architect can be handed over to specialists in some ambitious and prestigious buildings.

Interior designer
An interior designer is employed to design particular areas or plan individual fittings.

Landscape architect
Garden and landscape architects design and plan the exterior areas in agreement with the architect. Here, the landscape architect's brief can extend from creating ambitious private gardens to planning functional public squares or green spaces, sports grounds, or even noise protection facilities.

Facade planner
In the case of large, complex properties the architect can recommend that the client should employ a facade planner because of the diverse requirements of facades.

Lighting designer

Lighting designers or lighting planners can be commissioned to arrange particular and general lighting. They will simulate and plan the technical and creative effect by day and by night.

EXPERTS

Unlike specialist planners, experts do not provide specific planning. They act in an advisory capacity, prepare reports describing conditions or establishing causes, and suggesting solutions for problems arising. Experts can be brought in to deal with almost any area. The following are the most important fields of activity in building:

Soil experts
A soil expert or geologist may be needed, according to local conditions and subsoil, to provide information on possible foundation construction or existing groundwater, by means of trial digging and test drilling, or from existing maps.

> \\ Tip:
> It is worth setting up a meaningful project structure at an early stage. All participants needed in specialist planning roles must be brought in at the right time. Important elements of project organization include setting up an address list containing data for all participants, agreeing on regular discussion dates (jour fixe), drawing up written minutes with information about completing work with deadlines and agreements about data exchange arrangement between parties (DXF, DWG, PDF, etc.).

Building historians

If a building that is being refurbished has historical value, consulting a building historian can be beneficial. He or she will compile a history of the building and can offer assessments of structures worth preserving. This will at least establish the restoration horizon, i.e. the period of time within which the refurbishment should be performed.

Traffic planners

If the building project impinges on the local traffic situation or requires changes to the existing infrastructure and transport access, a traffic planner can be brought in.

Fire prevention experts

It is essential, especially in a building project that makes heavy demands on planning, to consult fire prevention experts. They can provide crucial planning information that conforms with the law and is likely to qualify for the required permissions by drawing up fire protection reports or concepts and checking that they are correctly implemented.

Heat and sound insulation experts

It makes sense to commission heat and sound insulation experts for many types of building. They will deal with heat, damp and sound insulation requirements for new build, but can also assess faults and damage in existing buildings.

Acousticians

Acoustics are another aspect of building physics assessments. This aspect deals less with insulation for impact, airborne and footfall sound than with calculating the best acoustics for demanding spaces, such as lecture theatres or concert halls. Here, acousticians are essential contacts for architects at the planning stage.

Pollution experts

Advice from pollution experts may be needed for existing buildings in particular, i.e. for conversion, refurbishment and redevelopment. They can examine and assess the construction materials already present in buildings. Current findings show that materials that can impair the wellbeing of occupants and users have regularly been used. Particular problems can be caused by effects on health during refurbishment and when removing harmful materials (e.g. asbestos).

Pollution experts' findings and suggestions are particularly important for correct tendering and the safe handling of hazardous materials.

Health and safety coordinators

EU building regulations insist that health and safety coordinators are employed once a building site exceeds a given size. This service can be performed by the client or the architect if they have the requisite qualifications, or by a separate person. This means that the building project will be monitored in terms of safety at the planning and realization stages, so that the client's and the contractor's employees and non-participating third parties are protected from danger as far as possible.

DEPARTMENTS AND AUTHORITIES

So far, the experts mentioned are only those involved in the planning stage of the project. The institutions and authorities whose permission is required at various stages must also be mentioned, as the project cannot be realized without them. For example, it makes sense even at the early stages of the project to mark out the general conditions for realizing the project using existing development plans provided by the municipal or local authorities.

Building control department

It is essentially the architect who will deal with the building authorities in the course of the project. They are responsible within the building control process (planning permission) for ensuring conformity with building and other relevant regulations.

Other departments may become involved in the process according to the scale and demands of the permissions sought. These may include the land registry and surveying department, the land holdings office, the town planning department, the monument preservation authorities, the environmental department, the civil engineering department, the city parks authority, and many others. For public building projects, plans usually have to be submitted to the local fire brigade or to a fire and disaster prevention office, who will provide information and define requirements for fire and rescue. › Chapter Planning process, Gaining permission

CONTRACTORS

Contractors are of course important partners in a building project, as well as the planners and relevant authorities. The tradespeople and building firms are the people who implement planning on site, using realization plans and service descriptions. Two different models are essentially available to a commissioning client:

Single service

The client can award service provision for individual trade services to firms able to provide the service concerned. (A trade service means work that is generally provided by one craft branch.)

General contractor

Alternatively, the client can award the entire contract to a single contractor, also known as a general contractor, who will carry out the work him-/herself on the basis of the plans provided, or employ subcontractors.
› Chapter Planning process, Tendering, Construction

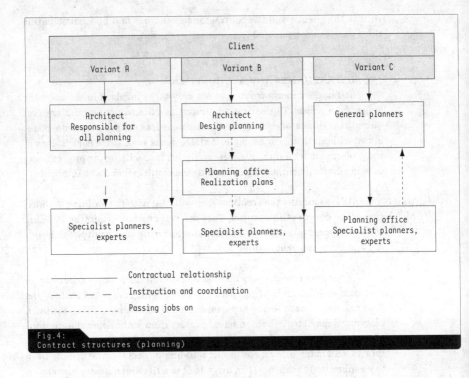

Fig. 4:
Contract structures (planning)

CONTRACT STRUCTURES

A client has various possibilities available for fixing the commissioning of planning and realization services contractually. › Fig. 4.

Planning contracts

_ The client concludes a separate contract with each planner and expert. Here the architect coordinates those involved in the planning.
_ The client enters into a contract with an architect to provide design services. Subsequent planning services relating to tendering and site management are transferred to specialist planning practices, which can also provide the required expertise and appropriate specialist planning.
_ The client concludes a contract with a general planner who will then commission all the other planners and experts needed for the project as subcontractors. Thus, the client has only one contact and contract partner for the entire planning process. Here the subcontractor undertakes, on behalf of another contractor (main contractor), to provide part of the service that main contractor has to supply for his or her client.

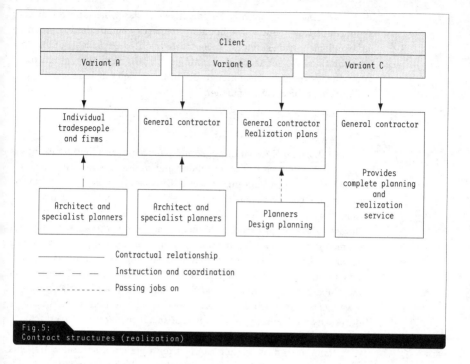

Fig. 5:
Contract structures (realization)

Realization contracts

Variants for commissioning realization services: › Fig. 5

- After general and detailed planning, the client concludes individual contracts with all the trades required. Here the architect will coordinate the individual trades.
- After general and detailed planning, the client concludes a contract with a main contractor who will provide all the services, or employ subcontractors.
- After submitting the design, the client concludes a contract with a general contractor who provides the engineering services that are still required (final planning, structural planning, expert planning, expert opinions, etc.) as well as the building services, or employs subcontractors.
- The client concludes a contract with a general contractor who takes over all the planning, including the design and subsequent realization, and employs all the other participants (architects, planners, experts, firms carrying out work, etc.) as subcontractors.

The ways of grouping the commissions described above can vary from country to country. The boundary between commissioning planning and implementation services can be drawn in different places. But for most combinations the sequence of planning steps will be similar, as the planning phases required have essentially to be carried out independently of the nature of the contract partner.

TEAM BUILDING

Cooperation

All those involved go through an intensive process in various phases from the first contact between client and architect until the keys are handed over when a building project is finished. It is important from the outset that a basis for cooperation is found so as to deal with critical situations that may arise during the project. Since the client has a justifiable interest in the project's being implemented within the agreed financial parameters and timescale, this does not mean a superficially harmonious cooperation, but involves all participants working constructively together to carry out the work at hand to their mutual satisfaction. Building can never be a covert activity, and so the planners and authorities involved, as well as the client and financier, represent a general interest that can be addressed by all concerned in a spirit of social cooperation. › Fig. 6

Planning team

Planning and realization can last for a period of several years. This depends on the size of the project, which can be dealt with in several building phases if the parameters require. So the planning team and also the underlying concept, the basic idea, should be viable in the face of all kinds of

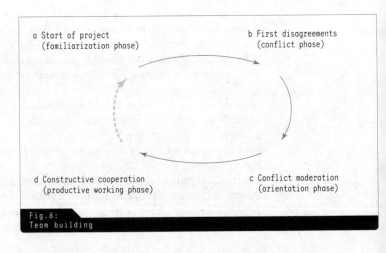

Fig. 6:
Team building

situations and challenges. A project can be planned in considerable detail and with great foresight, but as it proceeds new insights and changes will occur that need joint answers. As the project progresses, more and more planners become involved, the responsible authorities and departments have to be included, and increasing numbers of firms and tradespeople will be drawn into the work as services are put out to tender and commissioned.

Team ability

The roles that crop up in every project (client, architect, specialist planner, tradesperson, etc.) are usually played by changing partners and individuals. This makes the architect particularly significant as coordinator in the team-building process. The project team – like every other working group – goes through various social action phases.

The initial phase is often marked by expectant politeness, the team members tend to be excited, curious and waiting to get to know each another better. But the project at hand requires everyone involved to produce work that affects the others. This can lead to both professional and personal conflicts. If there are confrontations and tension, people should never lose their objectivity, and here the architect can be in demand as a mediator as well as a coordinator. It is necessary in this "orientation phase" to reach the mutual understanding that everyone is working towards the same goal, and this can be achieved only by working together and maintaining respectful forms of interaction and behavior. But confrontations over matters of expertise must definitely not be excluded.

Once the best working basis has been found, the team members should ideally approach each other in a spirit of trust, frankness, inventiveness and solidarity. This will make the project team able to work effectively, powerfully and purposefully towards realizing the project aim. So directing the planning team with this end clearly in sight is an essential part of the architect's work along with effective project management, and without it the planning team may lose sight of its goals.

GENERAL CONDITIONS AND AIMS

Planners have to deal with three basic targets at every project planning phase. These are crucially important to the client, but also to the architect. First come the cost framework defined by the client, and the time the building is due to be completed. The client will also express wishes in terms of built quality, which depends directly on the other two parameters, costing and scheduling.

COSTS

Building costs are enormously important to clients, and so they will expect reliable professional advice from the architect from the outset.

Cost security

The client derives cost security implications from the aims the building project is required to meet. Commercial, public and private building clients will differ in the way they formulate their aims.

A commercial client will address economic aims when determining a costing framework. Here the desired cost-effectiveness for their capital, or that of third parties, is most important. Both the continuing yield and retention of value for the future are important for this investment.

A private client is investing in his or her own future, as they want to use the property themselves. The important thing here is the future potential of the planned building project. Both the client's financial resources and stable value should be considered when implementing the planned project.

A public client's building intentions derive from the duty to provide the infrastructure for the services they have to deliver. Here the work will be differentiated according to administrative size and structure. For example, building commissions from regions or local communities can relate to

\\ Note:
A detailed treatment of this field can be found in *Thema: Baukosten- und Terminplanung* by Bert Bielefeld und Thomas Feuerabend, Birkhäuser Publishers 2007.

Fig. 7:
Establishing costs – cost specification

securing emergency services, fire and disaster protection, health services (hospitals) and education (schools). Smaller communities are responsible for providing kindergartens, adult education facilities, community centers, and other educational provision. A public client's decision will tend to be based on estimated need as well as social or idealistic aims. The cost-effectiveness of the measure will be defined by the longest possible period of use. But the resources available for such measures are restricted even for public clients, and subject to approval by the appropriate political committees.

Cost framework

A client can provide the architect with a cost estimate as a framework for implementing the project. This can be agreed as a binding upper limit or as an approximate target. The architect calculates feasibility on this basis, and provides the client with information about possible standards, floor area and volume on the basis of the budget. Here, imponderables and uncertainties in the nature and scope of the costs should be indicated.

Another possibility is for the client to supply the architect with information about function, spatial program and the desired quality of the built product. The architect will then estimate the expected costs on the basis of these requirements.

In practice, the two procedures are often mixed. As clients are usually accustomed to being certain about costs from other areas and business fields, keeping within the costing framework becomes increasingly significant even at an early stage.

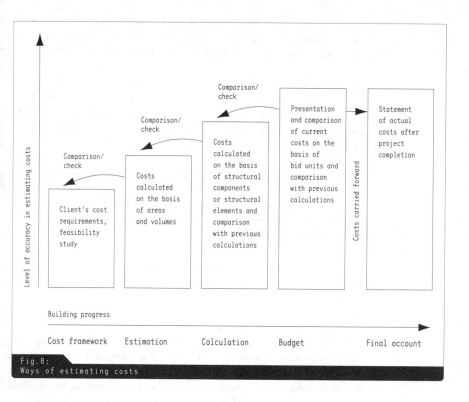

Fig. 8:
Ways of estimating costs

Cost specification

The architect will use a variety of instruments for estimating costs independently of the actual planning phase, ascertaining all costs as the product of a quantity factor and a cost specification. The cost specification reflects the ratio of costs to a reference unit like area or volume. For example, a typical cost specification would be EUR 1000/m² gross floor area. › Fig. 7

Steps in ascertaining costs

It is essential that the current cost development levels are available to the client as an aid to decision-making when planning stages require it. The first step is the decision to make a start on the project, to decide whether the idea actually is viable. And after the first sketches and drafts have been drawn up, the probable costs are estimated on the basis of very rough reference points. These <u>cost estimation</u> reference points can include the required useful floor space, the gross floor area or the gross volume of the property. If planning has moved on further and the design has been fixed, costs have to be calculated before a decision can be taken about submitting a building proposal. › Fig. 8

A more precise <u>cost calculation</u> can be drawn up on the basis of detailed information about the supporting structure and individual structural components, based on rough or more precise structural elements.

Rough elements might include exterior and interior walls, ceilings or roofs, and more precise elements the individual layers within the rough elements (ceiling rendering, reinforced concrete ceiling, screed, floor covering, etc.). The overall costs to be determined at this stage ensure a higher degree of certainty about costs than previous costings.

When services are being allotted to contractors and traders the next step is to present the current state of costs. This has to be reconciled with the previously established budget for the individual awards, in order to compare real market prices with the <u>cost planning</u> and then intervene to impose controls where necessary.

When the building project is complete the architect compiles the overall construction costs. This <u>cost determination</u> records the actual building costs accrued.

Cost control

It is important to check new findings and detailing in the cost planning against previous stages and the client's cost stipulations regularly, so that any discrepancies can be recognized at an early stage. The further a project progresses, the more difficult it is for the architect to intervene in cost development and bring discrepancies under control. In the building phase it is necessary to make mixed cost control calculations based on calculated costs, on services that have already been commissioned, and sometimes on services that have already been rendered and paid for, as different pieces of work may not run in phase. › Chapter Planning process, Construction

Cost tracking

Architects can build up their clients' trust by supplying an uninterrupted flow of information and by tracking costs conscientiously in

> \\ Important:
> The client has the right to demand a current statement of costs throughout the duration of the project. As the architect is interested in keeping to the cost stipulations formulated at the outset, he or she must always be in a position to list costs comprehensibly and present them so that they can be compared with previous cost information.

accordance with the original budget estimates, thus showing their clients that they are able to handle capital entrusted to them.

DEADLINES

Client's deadline stipulations

Like cost stipulations, deadline stipulations are important in many respects for the cost of the project. Notice may already have been given on existing rental contracts because long notice is needed, or the moving-in date may already have been fixed because of financial constraints (production starting, seasonal business, etc.). The architect has to address the stipulations critically in terms of their feasibility and plan the sequence of deadlines for the project in agreement with the client. Here it is not just the building phase that has to be taken into account, but also the preliminary planning.

The stipulated deadline will usually be based on the <u>final deadline</u> for realization. It is also possible to lay down a deadline for the <u>start of building</u>, if for example this is linked with the funding payments. Another variant on a time stipulation is the <u>shortest possible realization period</u>. Here the beginning and end of the process do not have to be laid down rigidly. A <u>short realization period</u> should reduce the restrictions and obstacles caused to the project by the building phase as far as possible.

Scope of the schedule

A <u>project schedule</u> is drawn up to represent the overall time needed to complete the project. This considers both the planning and the realization phase. It can help to compile a separate <u>production schedule</u> (work schedule) for the building phase for the sake of clarity. But here the fact that building depends on planning and commissioning contractors must always be taken into consideration. › Fig. 9

Planning the planning

At the beginning of the project, it will take a certain amount of time for the project to be clarified by the client and the architect. A complex process of agreement will then be developed involving the specialist planners and the responsible authorities. The architect has to establish a basis for work and circulate it to all concerned so that the planners can start work. Conversely, the architect is dependent on the planners' progress and their submission of completed documentation as the project proceeds. › Fig. 10

Forms of presentation

Today, schedules are worked out using dedicated computer programs. The simplest way of presenting the schedule can be a list naming the procedure and the start and end dates. But this is not particularly clear given the large number of processes involved in building projects.

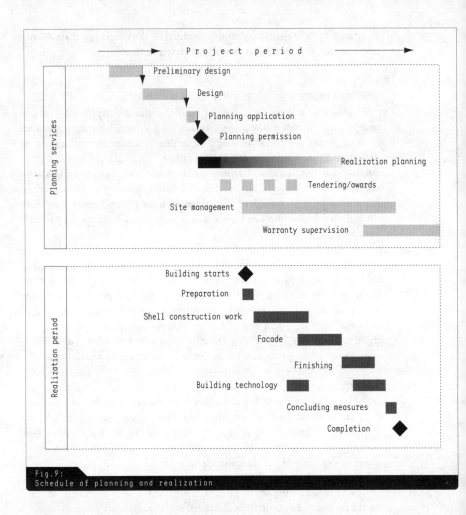

Fig.9:
Schedule of planning and realization

Network plan

A network plan shows the various processes in terms of their mutual dependency and the connections between them. This approach works well for illustrating the interlinking connections, but does not show time sequences.

Bar chart

The most usual presentation mode is the bar chart or Gantt diagram. The Y axis lists the individual events or processes and the X axis shows time. The duration of each process is represented by bars above the time axis. Overlaps and related items can be established without difficulty, especially with modern scheduling programs. Vital connections

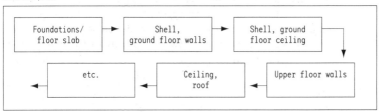

Fig.10:
Ways of representing schedule plans

can be shown by linking the end of process A → beginning of process B, etc.

Milestones

Milestones are important phases or direct deadline stipulations within the overall project schedule (building starts, completion, etc.). For example, building permission is an important scheduling milestone. Adequate time must be allowed for the examination of documents by the authorities, as necessary changes defined in the permission documents can affect the start of building. Implementing milestones and presenting events and processes in color make the bar chart clearer and more intelligible to all concerned.

Target-performance comparison

A target-performance comparison can be used to check the schedule against actual building progress to establish whether the project is running according to plan. The aim of scheduling is to identify at an early stage if the final deadline is at risk and to counter this where appropriate. It is only by constantly checking and monitoring processes that it is possible to get things back in hand in time.

QUALITY

Technical standards

In principle, quality can be defined in two different ways. Generally speaking, the current standards, directives, regulations and laws are an unalterable basis for planning and realization. These <u>codes of practice</u> are universally acknowledged among experts. Fundamentally they apply to every building project as a technical standard and do not have to be required or stated explicitly in contracts. However, correctly and comprehensively applying all valid standards and directives is an important sign of quality in a building project, and one that cannot be taken for granted at the realization stage.
> Fig. 11

Standard of individual finish

The client lays down the standard of individual finish for a building personally. He or she defines the building's visual appearance, describing all the structural components in terms of material quality, form and color, within the bounds of what is technically possible. Here the client's financial latitude will be one of the factors in determining the final standard, as there can be considerable cost differences for different standards of finish. The color scheme and the shape of the building usually make less impact on the budget than the choice of materials. For example, large areas of glass

Fig.11: International standardization organizations

Fig.12: Establishing finishing standards individually using samples

in the facade or the interior cost more than closed masonry or reinforced concrete surfaces. Using oak rather than plastic for windows is visually more prestigious, but more expensive; as it is to install high-quality parquet that needs special laying as opposed to a textile floor covering. These examples can be continued throughout the list of structural components. It is important for the architect to discuss the standard of finish with the client at an early stage, to help him or her to decide by presenting samples and to make it clear what the cost implications will be. › Fig. 12

Quality descriptions

The recognized codes of practice, directives and standards cannot be further defined in terms of quality. Individual standards of finish need to be clearly defined in writing, however: at the start of the project a non-binding standard of finish is fixed, e.g. "high standard" or "average standard", which leaves a great deal of scope for interpretation. Such statements are consistently refined in the course of the planning process.

First, the finish for the loadbearing structure and the essential building components has to be fixed. As the project proceeds, detailed decisions have to be taken about every component. The architect helps the client to decide by providing samples and finish variants with costings. Once the decisions have been taken, they are recorded by using a general qualitative building description or by compiling a detailed room book, in which rooms used for the same purpose can be considered together. The data sheet gives the name of the room, its number and floor, and information about walls, ceilings, floors, doors, windows, radiators, and sanitary and electrical fittings.

This quality description forms the basis for cost calculation and points to be fixed in the service specifications that will need be drawn up later. If the client expresses wishes about additions or changes in the course of the building process, the architect is required to point out the effect this will have on cost, with reference to the defined standard of finish.

Monitoring quality

The existing technical standards and the individually fixed standard of finish are monitored by the site manager during the building process. Here, quality control means checking materials when they are delivered and installed, monitoring the quality of craftsmanship, establishing that details are being executed as prescribed, and ensuring conformity with all current standards and laws. Consideration for legally defined dimensional tolerances is also part of this monitoring process.

The quality of a new building is reflected not just in a convincing design, but also in the correct application of technical standards and successfully implementing the client's individual requirements.

PLANNING PROCESS

DECIDING ON A PROJECT

At the start of a project, a building client has to clear up a number of fundamental issues. First of all, the financial viability of the project has to be checked. If the location for the new building has not been fixed, a suitable plot of land must be found. The client will also fix the completion date and cost framework for the building project. In the past, these questions were essentially settled by the client him-/herself before the actual planning phase started. As the general conditions building clients are working under are becoming ever more complex, the architect can now be a helpful contact even at this early stage. › Fig. 13

Scheduling and costing

At the beginning of a project the work to be covered must be clarified, the client's ideas and intentions identified and checked for feasibility. This involves both the client's financial possibilities and also a realistic estimate of the time the project will take. So, for example, the architect has to test the client's ideas against specimen projects to see whether his or her ideas can be implemented at all, and to discuss a realistic approach with him or her where appropriate.

Consultation about the site

The architect can advise the client about choosing a plot or assessing one that is already available. He or she will give hints and appraisals with reference to position, character, surroundings and other conditions relating to location for the planned building project. Involvement in acquiring a site or advice on possible finance is not usually one of the essential services provided by an architect.

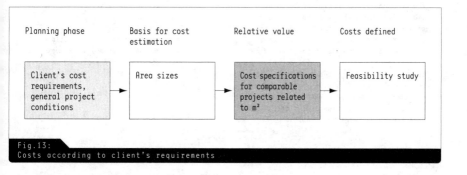

Fig.13:
Costs according to client's requirements

Authorities and specialist planners

An inexperienced client should then be informed about the necessary involvement of authorities and specialist planners. He or she has to know which additional planners are required to implement the building project and which additional reports, analyses and investigations have to be acquired or carried out before work can start, as in most cases he or she will have to conclude contracts with those concerned. > Chapter Project participants, Contract structures

Degree of planning involvement for participants

Not every client is familiar with the individual stages and development steps in the course of a building project. Architects are able to aid their clients in making decisions, especially when choosing the necessary specialist planners such as structural engineers and building technicians, and advise them about the specialists' competence and the scope and cost of specialist planning services; of course the architects' own services must be defined as well. Agreements of this kind should be made as early as possible, so that different views about the services commissioned and required do not emerge at a later stage.

>

Bidding for commissions and acquisition

At this early stage of a project, architects should support their clients' projects to the utmost and make it possible to realize the new building, as they are on the threshold between acquisition and commission. This exploratory phase is often seen as bidding for a commission without remuneration. Here it should be made clear to the potential client that the basic matters defined at the outset such as costing, timescale, choice of location, and selection of the people involved are crucial to the successful and conflict-free implementation of the building project.

>

\\ Note:
The appropriate fee tables or recommendations will show essential and additional services. Additional services could include models for demonstrating designs, detailed inventories, or drawing up room and function programs. Architects will do well to agree on the scope of the work phases to be commissioned, as well as on the service requirements associated with them and where appropriate any additional or special services that may need to be commissioned.

\\ Tip:
It is recommended that a written file memo should be created before every meeting with the client and distributed to all participants. This makes it possible to build on previously agreed decisions and subsequently confirms agreements hitherto made only verbally.

This first phase typically includes a high degree of consultation, so that the two most important partners in the project get to know each other and create a basis for successful and trusting cooperation.

Architect's contract/fee

> 🛈

Once a positive decision has been made about a project a written basis for a contract should be drawn up for the cooperation between client and architect, stating the scope of the work to be done and the fee the architect is to be paid.

CONCEPT PHASE

After the basis of the project has been defined and cooperation between the planners and the scope of their brief fixed contractually, the architect can start to address the actual conceptual building project at hand.

Sketches

The key data defined in writing now have to be translated into sketch form for the first time, in order to convey an idea of possible approaches to the project. > Fig. 14

Presenting the design

The way in which the design is presented is very largely left to the architect. He or she will usually make some early sketches by hand, or prepare simple CAD drawings to convey his or her ideas to the client. CAD systems in particular make it possible to create variants on views and ground plans with relatively little effort, to present the client with a selection of possible alternatives.

\\Note:
Architects are faced with a number of liability risks in the course of the planning process. Engineers and architects are threatened with liability not just at every realization phase, but also when the contract is concluded. They are liable not only for faulty planning, for mistakes in awarding tenders and for other infringements of contractual duties, but not infrequently for defects caused by the contractors as well. Most countries insist that professional risks indemnity insurance be taken out to cover these risks as far as possible.

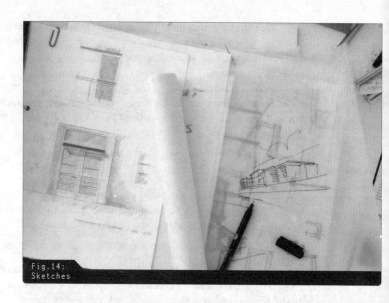

Fig.14:
Sketches

Fig.15:
Working model

Preliminary design

Ground plans, sections and views should be presented to scale, which will usually be 1:200 in the early stages.

The architect should always be mindful of making matters comprehensible for the client when presenting the preliminary design. He or she has had a great deal of practice in reading ground plans or section drawings. But someone who is not used to addressing plans on different scales will often see them as just an abstract representation, and may well need help in getting bearings and understanding the access situation and room disposition. The creative idea behind the plans, the three-dimensional effect or the urban connections with the surrounding area and the functional links between the differently used rooms and space usually need detailed exposition by the architect if he or she wishes to convince the client about the creative thinking invested. It is his or her job to work with the client on translating the latter's functional and financial stipulations into architecture. The client's suggestions and reservations can be extremely productive and stimulating for the architect's own work. So far it is not just the client who has been moving on unfamiliar terrain: the architect will also come across matters that are new when working on a variety of projects. › Fig. 15

› 🔎 A preliminary design that meets the requirements and that can convey a distinctive architectural approach or a creative idea will then crystallize out from exchanges with the client, the eventual user and the other specialist planners.

Cost estimation

Correct presentation to scale is also required at a relatively early stage, as it is on this basis that the client examines the way the required room planning is being implemented and decides about further work on the concept. A first project-related cost estimate will now also be drawn up as an additional basis for decisions. This involves establishing quantities via rough units (areas, volumes), and multiplying them by characteristic cost specifications for comparative objects. A cost specification represents

> 🔎
> \\ Example:
> An architectural practice that works only in housing construction will not usually need to readdress the current standards and directives for housing construction and the wishes expressed by housing associations. But for other building work, such as educational buildings (kindergartens, schools, colleges), sport (swimming pools, sports halls, etc.), buildings for cultural purposes (assembly areas, concert halls, stadiums), office or industrial complexes the architect will have to address particular project parameters such as special regulations for public buildings, workflow, etc.).

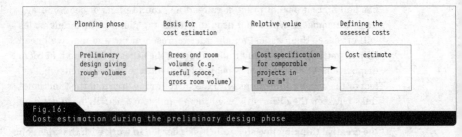

Fig.16:
Cost estimation during the preliminary design phase

the cost per quantity unit (e.g. EUR 1000/m² gross floor area). Cost specification levels are estimated on the basis of realized building projects of a similar character. As there are still a lot of imponderables at this early stage, the architect should make it clear to the client that the information will not be available in more precise form until a later stage, and that it presently represents the area in which the costings are currently moving.

> Chapter General conditions and aims, Costs > Fig. 16

DESIGN PHASE

After the client has decided to set a project in train, work starts on putting the concept into practice. Looking back at the jointly prepared results from the first concept phase should provide the following information as a basis:

_ Preliminary conceptual sketches (ground plans, views, sections, perspectives, etc.)
_ Function schemes
_ Space allocations
_ Distribution of quantities and areas

\\Tip:
As determining costs depends to large extent on volume and quality of finish, costs as presented are subject to a high uncertainty factor at this early stage. Cost risks can be considerably contained by a comparative costing in terms of gross volume, gross floor area, and useful space.

\\Example:
A building can have the same volume on a square or a long rectangular ground plan. But the comparatively expensive facade area will differ distinctly in ratio to the gross volume. The building with a square ground plan needs a considerably smaller facade area, which greatly reduces cost. Difference of this kind should be borne in mind when using cost specifications.

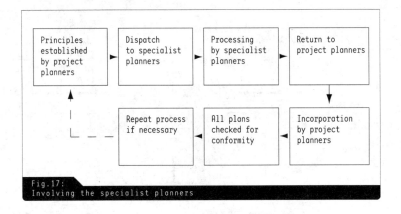

Fig. 17:
Involving the specialist planners

The design principle now needs further development on the basis of these requirements, concrete constructions should be fixed, the necessary technical fittings integrated, and the standards of finish agreed.

One concrete example: the position of a sports hall on the plot as been determined as part of the "conceptualization process". The required dimensions of the building are derived from the space needed for the playing area and the necessary ancillary rooms. The links between the individual spaces within the hall as a whole were established at the concept stage. The architect has already presented the first ideas about the external appearance of the building using sketches or views.

So far the type of loadbearing structure to span the hall, what heating or ventilation systems should be installed, where and in which outside walls apertures are required to provide light, and how many, and not least what materials will be used to build floor, wall and ceiling, have not yet been established.

System and integration planning

Thus there is a need for more discussions with the client and also with the specialist planners. The term "system and integration planning" is now used for this project phase. › Fig. 17

Various systems (loadbearing structure, heating, etc.) have to be discussed and coordinated; they cannot be considered separately because of the many interfaces needed. For example, if the necessary spaces needed for ventilation, lighting and installations are not considered in the case of a closed beam over the playing area, it is difficult to provide the necessary height for the hall economically. If the building technician is planning

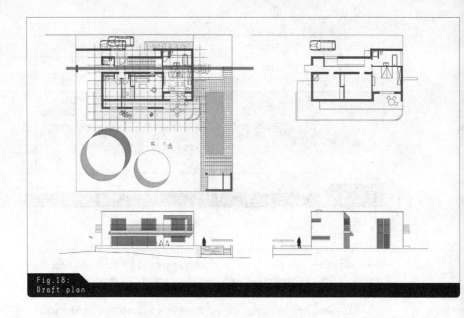

Fig.18:
Draft plan

the best possible heating equipment but forgets the planned large area of south-facing glazing, conditions could well be intolerable in the summer months. Hence the architect has to find the best possible solutions for the various systems, working with the specialist planners, and must define the mutual dependencies of the individual specialist plans in order to come up with the best overall solution.

Draft plans

Developing the preliminary design requires plans to be drawn up on a larger scale. All the ground plans, views and required sections are usually drawn on a scale of 1:100 in the next design phase. The aim is not to dimension all the sections of the building completely, but just to reproduce the essential external dimensions of the building, the relevant room dimensions, and where needed the aperture dimensions.

\\Note:
More precise information on plan presentation and dimensioning can be found in *Basics Technical Drawing* by Bert Bielefeld and Isabella Skiba, Birkhäuser Publishers 2007.

At this stage, a plan is not intended for working purposes, but is simply aimed at the client, the specialist planners, and the building authorities. It does not usually make sense to work in detail here, as the plan will change considerably in discussions with the client, the specialist planers and the authorities, and excessive detail would create an undue amount of work.

Plan presentation

Draft ground plans show room uses and room areas. The surrounding buildings and the existing topography should also be represented, according to locality.

The free presentation acceptable in the concept phase now has to give way to a largely uniform approach to drawing plans, as the completed draft plans also form the basis for the subsequent application for planning permission, and have to meet the authorities' requirements. Different countries have also issued their own presentation directives, though much is laid down in international standards.

Exchanging plans

The draft plan must provide the client with details about the form the building will take, and the architect also exchanges plans with the specialists involved. It therefore makes sense to clarify the interfaces between the planners and the data transfer type (e.g. DXF, DWG or PDF formats with a fixed level structure), so that information can be exchanged as efficiently as possible and without unnecessary conversions.

Quality of finish

The basic quality of finish and the standards for fittings have to be fixed with the client as part of the draft planning process. A detailed description of the planned building should be drawn up as a basis for later work by all those involved in the planning. It will be revised in subsequent planning steps recording all the qualities of finish that have been decided. This means that the architect can refer back to the originally defined standards if changes are requested later, and thus justify additional costs.

Cost calculation

More precise bases for planning and details on chosen systems (load-bearing structure, building technology, etc.) now make it possible to assess the costs more thoroughly. At this stage in the planning, building costs should no longer be assessed in terms of interfaces and building volume › Chapter General conditions and aims, Costs but structured in detail around individual cost groups such as walls, ceilings, roof, heating system, exterior areas, etc. This gives the client a serious basis for deciding whether the building project should go ahead as planned, which means that the plans can be submitted to the building authorities in their present state as an application for planning permission. › Fig. 19

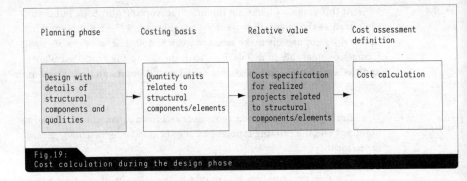

Fig.19:
Cost calculation during the design phase

If the costs are identified in such detail that they can be allotted to contractors (trades) at a later stage, this will make it possible to produce a targeted cost comparison in terms of previous cost estimates when awarding service contracts, thus meeting a key requirement for transparent cost monitoring.

When working out the design, the architect must produce plans capable of gaining planning permission. As these plans will form the basis for permission and for drawing up working plans, prior negotiations with the authorities may be needed to ensure that they are viable for such submission. Here a major part is played by current legal regulations and standards, which can vary considerably from region to region. Their validity is also affected by the function to be accommodated (housing, work, places for assembly, etc.). The main requirements to be met relate to fire prevention, protection at work, places for assembly, housing construction standards, etc.

These provisions are there to protect the user, the environment and society as a whole. The architect has to advise the client about legal requirements and observing the rules as part of this design phase.

GAINING PERMISSION

Application for planning permission

The responsible authority now has to decide whether the planned project will qualify for planning permission. The essential information is already available in the scaled plans that have been drawn up. All that may need to be added are dimension chains, room names, and information about areas and volumes. The surveyor who will have been involved from an early stage completes a site plan, unless this has already been prepared by the architect or the building authorities. The architect is responsible for compiling or revising all existing plans in an orderly manner, providing a uniform plan heading and signature line for the client. Some authorities require this signature on the plans. It is also an important component of the architect's contractual relationship with his or her client. The subsequent working plans will be drawn up on the basis of these first plans. If the client wants changes to be made at later stages, the architect will be able to refer to the jointly agreed design or permission plans where needed. Desired changes often result in additional costs or delays in the building phase, which can be documented comprehensibly on the planning basis that has already been established.

Forms and documents

Another set of forms has to be submitted along with the planning documents, giving the authorities structured information about the client and the key people involved at the planning stage such as the architect and structural engineer. The building project is also described in words, the building type defined (housing, school, etc.), as well as the type of use, estimated building costs, net floor area, enclosed space, and the essential building materials to be used.

The structural engineer records his or her qualifications and the statical information relevant to the building. According to the country, proof that energy-saving regulations have been met may be required, and according to the nature and size of the project, a sound insulation and fire prevention certificate, a certificate of the required number of parking spaces, and the site ground use ratio. To secure access, an additional drainage application is usually required, showing connections to local sewerage systems and public utility supplies.

The authorities now check the plans submitted with reference to the building regulations, but not for possible structural or functional defects. Even incorrect information from the authorities about the proper implementation of all the valid regulations does not protect the architect from possible liability claims. Architects are therefore well advised to make themselves entirely familiar with all the relevant laws and locally applicable building regulations.

Duty to advise and inform

If doubts arise about whether a client's wishes are likely to obtain permission, the architect should point this out to the client at an early stage and will recommend that experienced legal advisers be brought in where necessary. Architects should not allow themselves to be drawn into giving explicit legal advice in controversial cases, although they do have a general duty to advise and instruct their employers.

Submitting documents

It is usually the architect's duty to submit the entire package of documents. Requirements about the form of the necessary documents vary nationally and regionally. It therefore makes sense to contact the authorities at an early stage to clarify the formal requirements.

Once the examination has been completed, written planning permission will be sent directly to the client. As a rule, this is accompanied by detailed information from the individual departments about the standards that have to be met. If there are still discrepancies in the documents, such as failure to identify the site manager responsible, or omission of the structural engineer's certificate of competence, these details can be submitted subsequently.

Information and conditions

The planning permission will also contain information and conditions for the realization phase, e.g. with reference to:

_ Emergency escape routes
_ Appropriate provision for the handicapped
_ Statutory labor protection requirements
_ Fire prevention
_ Tree protection
_ Immission protection
_ etc.

All these requirements must be observed and implemented at the realization phase, as must the requirement that the project must be inspected by the building authorities in an on-site visit before the building is used. Here, the authorities will check the building with the architect and the client on site after completion on the basis of the application for planning permission and the conditions and comments made by the authorities.

\\Tip:
It is also now established practice to submit a sufficient number of copies of each document. An authority will inevitable take time to examine documents, especially if they have to pass through different departments (fire brigade, transport office, trading standards department, etc.). If sufficient copies of applications for permission and planning documents are submitted, they can be distributed to all the departments involved at the same time, which will speed up their return considerably.

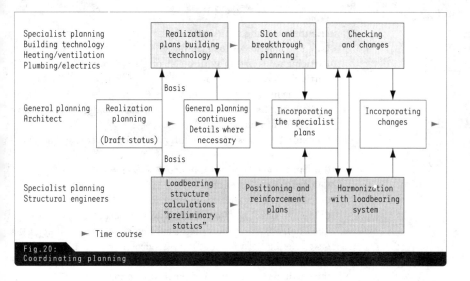

Fig.20:
Coordinating planning

WORKING PLANS AND QUALITY OF FINISH

The previous working phases on the project were aimed at clarifying and fixing the client's ideas. The detailed plans produced in this way have been checked with reference to the building authorities for their viability in the permission process and the feasibility of the building project.

Bringing in the specialist planners

The realization phase is central to further processing of the existing plans. The participating specialist planners are now key figures in the process. In concrete terms, the architect must inform all the specialist planners about the current state of the plans, and pass on the building authorities' conditions to them. The structural engineers can compile a preliminary statics report showing the main dimensions of the loadbearing structural components on the basis of the draft plans, if this has not already been done at the preliminary planning stage. The building and electrical engineers need information relating to fire protection (fire alarms, fire compartments), requirements for drainage and connection to the local supply systems like water, sewerage, gas, electricity, and data cables.

Planning meetings

It makes sense to hold regular planning meetings at intervals to be agreed jointly. A pattern that seems to work is a fixed date on the architect's or the client's premises, on a weekly cycle. The architect keeps minutes or notes of the meetings, containing the decisions that have been taken. The contents of minutes can be structured as follows:

Structuring minutes

- Project title
- Minutes numbered consecutively
- Date and location of the meeting
- Participants (starting with a list of participants, in which each participant records his or her name and contact details such as telephone number and e-mail address)
- Distribution list (all the participants present and people who have to be informed additionally)
- Content of minutes. Here the following tabular structure seems to work:
  Column 1: Agenda points numbered consecutively
  Column 2: Heading with agenda point followed by content
  Column 3: Responsibility (who has to deal with what)
  Column 4: Deadlines for dealing with this part of the project
  The final, obligatory point of discussion should be the date of the next meeting, location and the attendees required. Items for discussion that are already established should be identified, so that the participants can come to the next meeting prepared.
- The minutes will be signed by the person who has compiled them and any documents attached listed (schedules, planning documents, etc.).

There are of course various views about how minutes should be organized. The aim is to record the content of the discussions briefly and succinctly, thus creating a basis for further meetings and following up the decisions taken.

Client's participation

It also makes sense at this stage for the client to be involved in the meetings and for all the minutes and other essential correspondence between the planners to be passed on to the client. The client has to make a large number of decisions about realization details. The planners will provide the necessary documentation for this, and will have to advise him or her about the consequences in terms of cost, schedules, and the quality of finish.

Coordination

Constructive cooperation and good coordination between the individual specialist planners is an important prerequisite for the building site to run smoothly. Meetings of this kind will continue into the building phases, when site meetings will take place, as well as planning meetings.

Planning sequence

It therefore makes sense to agree on a planning sequence as part of the scheduling process. The structural engineer, the electrical engineer and the building technology engineer take the architect's plans as the basis for their own planning. Their planning work is then built into the architect's working plans. Thus each is partly dependent on the other, and also needs a certain amount of time to deliver his or her own part of the work appropriately.

Specialist planner's work

The architect fixes the dimensions and details for the loadbearing structural components in cooperation with the structural engineer. Construction variants are examined in terms of their financial viability. The building services engineer plans the heating system, the necessary cooling and ventilation systems, the plumbing required, and possible alternative energy sources. The electrical engineer defines the electricity supply standard, the necessary data cabling, and the lighting concept. If other specialist planners are involved, such as an acoustician, a landscape architect or an interior designer, they will also ask the client for information about the work they are required to do. And not least, the architect him-/herself has to make a number of decisions about materials, colors and shapes for all structural components.

Agreement between specialist planners

The way the different disciplines interlock and the necessity for agreement at the earliest possible stage will be illustrated taking additional floor height as an example.

Even at the design stage the client will have identified the height for the rooms. The story or construction height has been defined on the basis of local conditions and the maximum possible building height, as part of the permission-seeking process. Now the client would like to fit a cabled power and data supply via floor ducts, with supply sockets in the floor. At the same time, the heating engineer responsible is planning to install underfloor heating. This will mean that it is scarcely possible to realize the room height as originally planned, as building the necessary ducts and floor heating into the space under the floor will increase its height significantly.

> \\ Note:
> Keeping the planning and site meetings separate is strongly recommended. Good results are not achieved if there are disputes about planning details in the presence of the people doing the work that the planners could have agreed some time ago.

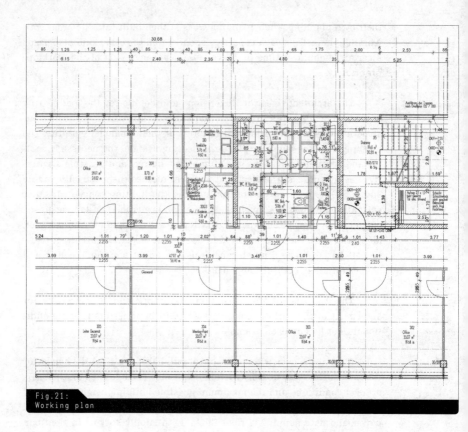

Fig. 21:
Working plan

Information flow

This example shows that different requirements can be formulated that may well not be met within the planned dimension parameters. Solutions can usually be found for problems of this kind by including the planners, to ensure discussion between them and a seamless flow of information. The earlier and the more thoroughly the planning takes place, the less the subsequent planning and building sequence is likely to be disrupted. Increased costs and extended building periods often arise as a result of superficial and imprecise planning. › Fig. 21

Working plans

Project plans will now usually be drawn in detail on a scale of 1:50 or smaller. These so-called "working plans" for all the trades are complemented by the necessary details on an appropriate scale.

Data exchange

As most drawing today uses CAD systems, data will usually be exchanged in digital form. Standards have been established for the date exchange formats currently on the market. Correct vector data are needed for further processing by a planning partner (e.g. DWG or DXF file formats). These data do reflect the precise dimensions of a building, but they are often presented in very different forms, as the different specialists work with very different CAD systems (architecture, statics, building services, surveying). It may thus be advantageous to send printed plans, called pixel files (TIFF, JPG, PDF), as well as exchanging vector data.

Plan contents

The approach and style for building realization given below should be used in dealing with plan content. Clear structuring into ground plans, views and all the necessary sections is essential. Logical presentation of the details and consecutive numbering for all planning documents require that the sets of plans be structured for the trades doing the work as well.

TENDERING

Tendering documents

After dealing with the permission process and drawing up working plans, the architect then has to compile the documents for inviting bids to carry out the building work. Information previously drawn on the plans now has to be clearly structured in written form.

If a tradesperson is to submit a bid that is realistic in market terms, he or she needs full information about the work required. This information is supplied in the form of tendering documents, which contain general information about the building project, contractual conditions, technical requirements, a description of the specific work required, and explanatory plans.

\\Tip:
It is not necessary to present sections of a building that can be covered by drawings multiply and separately in detail for each trade. A detailed section through an attic or the edge of a roof, for example, can contain information needed by the shell constructor, the carpenter, the roofer, and the plumber. The shell builder will discover the thickness and edging needed for the attic upstand. The carpenter recognizes the position and dimensions of the purlins, the roofer the form of the roof projection, and the plumber the desired finish for the flashing, sheet metal or guttering required.

\\Example:
Some trades, such as facade constructors, will need a complete detail series for the various detailed points. A sound basis for tendering and the subsequent execution of this work can be provided by using a clear structure with reference to top, bottom and side facade connections, logically numbered.

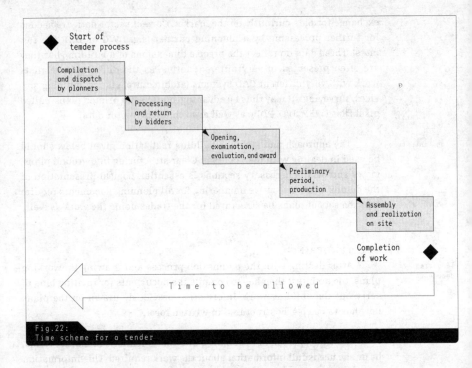

Fig.22:
Time scheme for a tender

Qualities/costs/deadlines

Qualities and standards of finish are defined within descriptions of the work required. The planner must take the available budget into account when describing the work. When the various commissions are awarded, the client will have a chance for a first check of actual market prices as compared with the costs established by the architect.

When planning and drawing up the tender documents the client's schedule requirements are significant in two respects. First, the architect has to have the tendering documents available at the right time for the

> \\ Note:
> Detailed information about bids and awards can be found in *Basics Tendering* by Tim Brandt and Sebastian Franssen, Birkhäuser Publishers 2007.

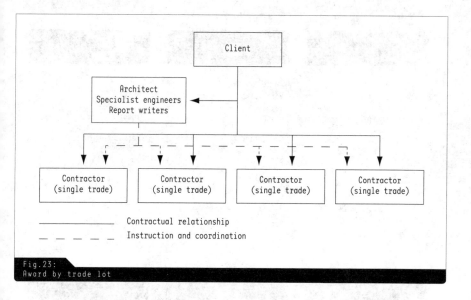

Fig. 23:
Award by trade lot

work to be awarded to contractors within the correct timescale, and that work can be started according to schedule. > Chapter Planning process, Tendering Second, the tender documents contain deadlines for the contractors for carrying out the work as part of the contract.

Tender scheduling

As part of the deadline monitoring process, the architect has to draw up and dispatch tendering documents so that the trades and contractors can be available at the right time for the building phase to run as planned. Sufficient time should be left for formulating the descriptions of the work required. The client's requirements must be taken into account, and possible overlaps with work by other specialist planners identified. Ambitious work may require agreement with specialist firms and advisers. Insight gained in this way can also affect the final planning stage or the way in which additional details are worked out.

When drawing up tender documents, the planner will take note of the time sequence at the building site. For building tenders of this kind the individual service packages or bidding units can be brought together in terms of time to reduce the amount of organization and editing needed. Or as an alternative to tendering by individual bidding units, the work required can be put out to tender in toto and awarded to a single contractor.
> Chapter Project participants, Contractors

Preparatory measures	Shell	Building envelope Facade
Site preparation	Earthworks	Roofing
Demolition work	Masonry	Plumbing
Earthworks	Concreting	Thermal insulation
Support works	Steelwork	Plastering
...	Sealing	Metalwork
	Joinery	Glazing
	Scaffolding	Painting
	...	...

Finishing	Building technology	Final work
Plastering	Heating installation	Cleaning the building
Screed work	Sanitary installations	Locks
Floor coverings in general	Ventilation	Signage
Tiling	Electrics	Outdoor areas
Dry construction	Elevators	Clearing the building site
Carpentry	Media technology	...
Painting	...	
...		

Fig. 24:
Typical trade distribution

If the tender accompanies the building process the work can be carried out in phases. But planning and tendering for an award of all the work for the project must be carried out before building starts, which means more time must be allowed for planning before building starts.
› Fig. 22

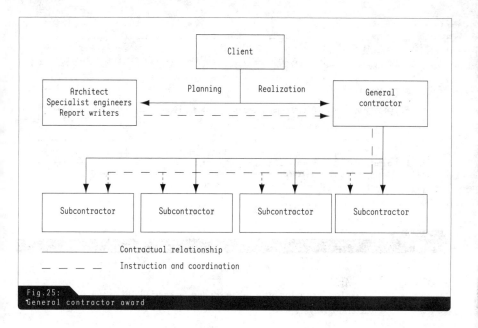

Fig. 25:
General contractor award

Tendering by trades (trade lots)

A trade defines the scope of a piece of work that can generally be performed by specialist craft or technical firms. The classical way of awarding building work is by offering individual tenders according to trade – a bid unit is the term defining the work to be done within a contract. According to the situation and the requirements, several trades may be combined in one bid unit (package award), or subdivided into smaller bid units (part lots).

Tendering by part lots

If the work required is very extensive (e.g. building a motorway), or if there are other grounds like risk distribution or including several firms for reasons of capacity (several building phases, working in parallel), then a trade service can be broken down into part lots. This is the term used when a trade service is split down into several sections requiring similar or identical craft services. › Fig. 25

General contractor awards

All the trade lots may be awarded to a single contractor (general contractor). The advantage here is that the person awarding the contract has only one contact, and only one contract partner, for carrying out the work. The general contractor is responsible for coordinating individual pieces of work. As a rule, a fixed price and a completion date are agreed with a general contractor. › Chapter Project participants, Contract structures

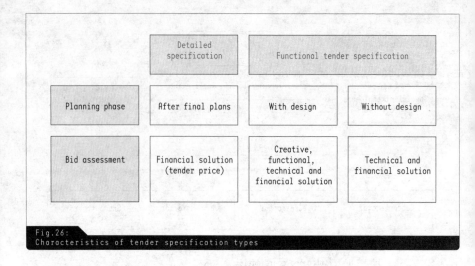

Fig.26:
Characteristics of tender specification types

Tender types

Tendering for building work can be carried out in two fundamentally different ways. › Fig. 26

Detailed tender (bill of quantities)

Detailed tendering by individual trades requires complete, thoroughly structured working and detailed plans. Individual tenders are then put out on this basis, according to defined award units. Both the required result and the procedure for carrying out the work are laid down in detail. The contractor finds a detailed specification giving quantities, materials and building techniques in the tender. The planner must describe the work required completely and unambiguously, so that each bidder is working on the same basis when calculating a bid. A tender specification is structured into the following components:

_ Title
_ Subtitle
_ Job item › Fig. 27

Structure level: title

The actual tender specification is broken down into titles dividing subsections of the work meaningfully. This grouping of subjobs by trade or room structures the tender specification, and it becomes easier to calculate and check prices (example: the roof as a bid unit can be divided into title substructure, title roof covering and title drainage).

```
Detailed tender specification
Trade: painting – interior

Title: 03 Painting – walls

─────────────────────────────────────────────────────────────
Item no.   Short text
           ..............................................................
           Long text                    Unit price        Total price
           ..............................................................
           Quantity - unit              (up)              (gp)
─────────────────────────────────────────────────────────────

03         Walls

03.10      Wall paint

           Check surface for suitability,
           support and adhesion qualities
           clean surfaces, prime absorbent surfaces.
           Intermediate and final coat
           Gloss undercoat: dull matt
           Color shade: old white


           Make: ........................
           (Information from bidder)

           350 m²
```

Fig.27:
Example of a tender item

Structure level: subtitle	Subtitles enable further title divisions. It can make sense in the case of large building projects in particular or complex trades to subdivide in this way, and thus establish distinctions within the job (for example by building phases or structural components).
Structure level: job items	The smallest subunit in a tender specification is the job item, which describes the work to be done. Here a distinction is made between short and long text.
Long text/ short text	Short text is the name for a job item heading. Under each heading is a long text describing the work to be done comprehensively and in a way that can be generally understood. Standard text models can be used, but they can also be formulated as wished. Figures are also provided for the required "quantity" and the "quantity unit", e.g. 25 m^3.

In order to provide correct information about the scope of the work, relating to each job item, the architect must establish the necessary quantities and dimensions (m, m², m³, 25 of each, etc.), using the existing working plans.

Functional tender specification (tender program)

A functional invitation to tender is formulated more generally and simply, and describes the desired result of the entire building project. It fixes a completion date, but leaves organizing the building work to the contractor as far as possible. As essentially only the desired aims are described, and some of the planning work devolves to the contractor, it is possible to award the work being tendered for at an earlier stage in the planning process. But the client and the architect have much less influence on implementation and the treatment of detail than in a detailed invitation to tender, which can have a deleterious effect on design quality. If this tender type is chosen, a decision has to be made between a tender specification with or without design.

A tender specification without design identifies the requirements made of the building on the basis of a construction or room program. The aim here is that an ideas competition should be held as the contractor draws up the tender documents. If a design is submitted, the structural and room program requirements are supplemented by concrete ideas about design and arrangement. But the functional nature of the tender is preserved, as the way in which the planning is implemented technically is left largely to the contractor. In a functional tender specification, the bidding contractor takes on the quantity risk, as he or she has to establish materials, building processes and quantities him-/herself and factor them into a bid prize. This is not the case with a detailed tender specification.

Structuring a tender

In principle a tender is structured in the same way whether it is in detail or program form. It will look like this: > Fig. 28

_ Textual elements
  _ General information about the project
  _ Contractual conditions
  _ Technical requirements
  _ Information about building site conditions
  _ Tender specification (detailed or functional)

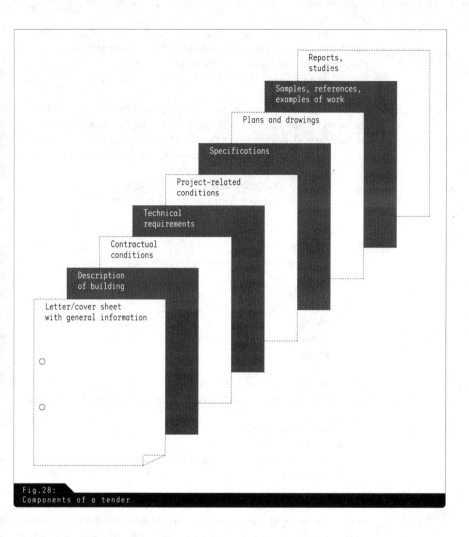

Fig. 28:
Components of a tender

- Drawn elements
  - Site plan/site mobilization plan
  - Ground plans / views / sections
  - Detailed plans if needed
- Other elements
  - Pictorial documentation
  - Samples
  - References/examples of finish
  - Reports/studies

General information about the project

The cover sheet lists general information and award modalities (submission deadline, realization deadlines, nature of award process, etc.). The client and the key individuals involved in planning are named. A short, general description of the project should also be placed at the beginning.

Contractual conditions

Once a commission has been agreed, the bidder's submission becomes the basis for the contract, working from the tender. Arrangements fixing the contractual modalities are therefore built into the tender documents. A distinction is made between general and special contractual conditions.

General contractual conditions are based on national and international standards, and include the following information:

- Nature and scope of the work
- Remuneration
- Deadlines for realization
- Notice
- Liability
- Building acceptance
- Warranties
- Invoicing
- Payments

Special contractual conditions are project related. They complement the general contract conditions and contain the following information:

- Submitting invoices
- Payment modalities
- Arrangements for subcontractors' services
- Discounts

Technical requirements

General and special technical requirements are seen to include standards in the sense of the recognized codes of practice and project-related, additional or higher demands made on the quality of the completed work.

Building site conditions

A description of the situation on the building site is important information, complementing the site mobilization plan, for the contractor when calculating his or her bid. It makes sense to include the following information:

- Location / address
- Possible road access
- Storage facilities

- Scaffolding / cranes
- Site-supplied water and power
- Sanitary facilities

Tender specification
The tender specification is at the heart of the tender documents. › see above

Drawing elements
The drawings and sketches enclosed are intended to help the bidder to fully understand the work required. It is perfectly acceptable to submit the planning documents in a reduced form. But representation to scale is urgently required to make it easier to calculate and check quantities using the plans.

Other descriptive elements
Any samples, photographs, etc. accompanying the tender documents are intended to complement the service descriptions outlined in the textual section and the plans.

Whatever kind of tender process is chosen by the client and the architect, the greater knowledge of the individual craftspeople is always needed. In this phase the architect wears his or her experienced "building services provider's" hat, rather than working as a creative designer. The tender is the interface and connection between planning and realization. Implementing high-quality design can only succeed if the tender phase is executed conscientiously.

THE AWARD PROCEDURE

After the architect has assembled the necessary documents in order to be able to attract tenders for carrying out the building process, he or she then has to find suitable firms and craftspeople who are in a position to realize the building project in line with the client's and the architect's ideas.

Suitable firms are those who have the necessary specialist knowledge and prove reliable enough to carry out the work required, have sufficient capacity in terms of personnel and machines available at the time the work is to be carried out, and can realize the project within the estimated cost framework.

These contractors, craftspeople or building firms can be found in various ways. For smaller, private building projects the architect can suggest to the client craftspeople with whom he or she has often worked before and had good experiences with them, so that a satisfactory result can be expected.

For larger projects, and especially for public construction works (national, regional, local authority), the architect is obliged to request bids for the work required on the basis of various procedural models. > Fig. 29

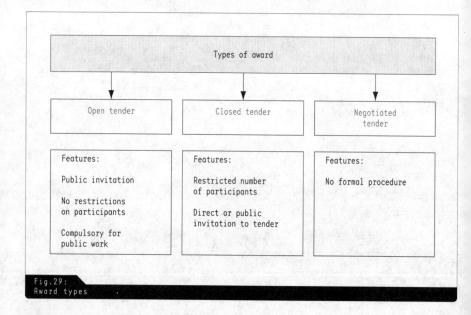

Fig. 29:
Award types

Open tender

For public building projects, tenders have to be issued and bids sought on a public basis. The bidders are informed about the tender procedure by announcement in the press, and can acquire the application documents from the client or the architect, against a fee. In open procedures of this kind, the lowest bid should be taken in unlimited competition. The basic principle of this kind of award is that there is a public invitation to submit a bid, and no restriction on the number of participants.

Closed tender

A restricted procedure may be used only if there is good reason, in exceptional circumstances. Here, a limited number of contractors are directly invited to submit a bid. Before issuing the tender documents it is important to ascertain that the selected firms are prepared in principle to submit a bid, otherwise there is a risk that no bids will be submitted before the opening date. Unlike a public tendering process, it is no longer permissible to investigate a bidder's professional competence after the bids have been opened. Merely by selecting the potential bidders to be approached the architect is bindingly confirming their expertise and financial and technical capabilities to the client, as well as their reliability.

The chosen bidders are sent the documents directly, without a fee, in contrast to the public tendering process. Opening the bids is identical in each case.

Negotiated tender

Negotiated or private award procedure must be specifically justified for public building projects. The required competition is restricted even further in such cases, as here the client negotiates with one bidder only.

› Fig. 30

> \\ Note:
> Of course a private client is not obliged to choose one of the processes described, and will thus probably tend not to keep to an open process or a negotiated procedure when awarding building contracts. But it still highly advisable to request several bids, in order to achieve a price in conformity with the market. Bid prices can vary considerably. A bidder will usually be more inclined to submit a more reasonable bid if aware that others are also bidding.

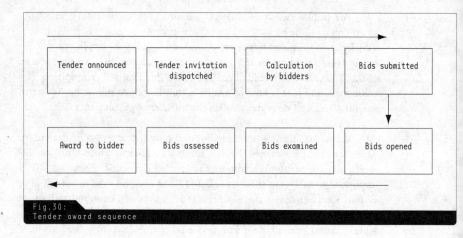

Fig.30:
Tender award sequence

Opening date/ submission

In public award procedures a binding submission date is named, by which all bids must have been submitted. Bids are opened at the appointed time, in the presence of bidders. Bids that are submitted later cannot be accepted because transparency and equality of opportunity are essential.

Computational and technical examination

Once the bids have been opened the architect must examine and assess them in detail. A formal examination establishes whether they are correctly signed and that nothing has been crossed out or added within the bid documents. This is essential if the bids are to be compared with each other. Otherwise a bidder could gain an unfair advantage vis-à-vis a competitor who has met the requirements correctly. The computational examination is concerned not just with comparing the bid prices; the key

\\Note:
The unit price is the bidder's statement of cost for one quantity unit within a service item. This forms the basis for invoicing the work actually delivered at a later stage. The total price is arrived at by multiplying the unit price by the stated quantity or the quantity actually billed. The net bid sum is the sum of all the total prices.

```
Invitation to tender for painting
Limited tender

Price comparison list

Spec.      Short text    Quantity/unit        Bidder A          Bidder B          Bidder C
item

1.10       Cleaning      50 m²                € 3.50            € 4.00            € 4.20
                                              € 175.00          € 200.00          € 225.00

1.20       Preliminary   50 m²                € 2.50            € 2.40            € 2.90
           coat                                € 125.00          € 120.00          € 145.00

1.30       Undercoat     100 m²               € 4.50            € 5.00            € 6.20
                                              € 450.00          € 500.00          € 620.00

1.40       Topcoat       100 m²               € 0.50            € 0.80            € 1.20
                                              € 50.00           € 80.00           € 120.00
  :          :              :                   :                 :                 :

Total      Title 1       net                  € 850.50          € 915.00          € 1280.20
           Title 2       net                  € 1320.00         € 1280.50         € 1450.90
           Title 3       net                  € 720.00          € 835.00          € 830.50
  :          :              :                   :                 :                 :

Total                    net                  € 4,850.50        € 5,210.50        € 6,160.30
                         gross                € 5,771.50        € 6,200.50        € 7,330.76

                         Discount             -                 -                 -
                         Deduction            -                 5%                -
                         Overall total        € 5,771.50        € 5,890.48        € 7,330.76
```

Fig. 31:
Price comparison

factor here is comparing the individual prices and the calculations made, as the unit prices will become the basis for the contract.

Price comparison   A price comparison list is drawn up in order to present the bids manageably. › Fig. 31 Here all the bid prices are inputted into a computer program for comparison. In principle, this could be done using a spreadsheet program like Excel, but tendering-award-invoicing programs are customary, which make it easier to compile a price comparison list. The minimum and maximum prices are already clearly marked here, and differences can

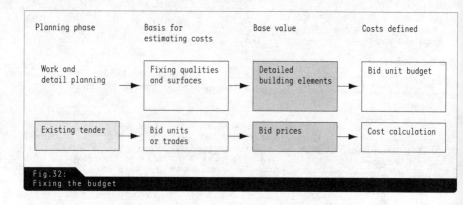

Fig. 32: Fixing the budget

be eliminated using percentages, and so deviations in unit prices between the bids can be identified. This is the first indication of the nature of the bidder's calculation and possible misunderstandings of the texts in the tender documents.

Award proposal   After finishing the examination, the architect submits a commissioning proposal to the client, which makes it possible to expect that the required work can be delivered within budget. > Fig. 32

Cost control   Once all the bids from the individual trades or bid units are available, it becomes possible to compare the calculated and the approved budget. This provides the client with reliable evidence about whether the architect has calculated the costs correctly or not. It is vital to compile the budget on the basis of bid units while cost calculations are still at the planning stage, so that costs can be readily understood. Then if tendering ensues on the basis of single trades or trade lots, it is possible to intervene and impose control on subsequent tenders if costs are exceeded.

Award meeting   Before the client commissions a contractor, it may be advisable to hold an award meeting. In public building projects, later negotiation of bid prices is prohibited, but for a private client it can be an instrument for establishing favorable prices. An award meeting can be useful as part of a public tender procedure in clearing up questions about realization effectiveness and possible alternative offers.

Building contract   The actual building contract between client and contractor or craftsperson is not primarily a matter for the architect, although he or she does have to advise the client about contractual items such as deadlines, contract penalties, payment modalities, discounts, warranty periods, safety retentions or similar matters.

///Time required

Reference has been made above to the time needed to draw up tender documents correctly. > Chapter Planning process, Tendering Sufficient time must also be planned into scheduling of the bidding procedure.

For a public building project, legal deadlines have to be met as part of the award process. After the tender specification for work required has been published and contractors have applied for the documents, the bidders need sufficient time to compile their offers. After these have been returned, the planner also needs a certain amount of time to assess the offers, so that they can be examined conscientiously. If the award procedure for a public contract is to be conducted by political committees, times for meetings should be taken into account, as well as deadlines for drawing up the documents needed for those meetings. Thus the period of time between drawing up the tender specification and commissioning a contractor can extend to several months, according to the type of client. Some trades needing to prepare workshop drawings and to order and process materials need a long lead time before doing the work on the building site; this should also be considered in the scheduling context.

CONSTRUCTION

All the planning activities described so far are needed to arrive at the actual core of the project: implementing the plans on the building site.

///Start of building

In public projects, the actual start of building is a good enough reason for the client to celebrate, as it is now that the public and the neighborhood will actually acknowledge in earnest that things really are under way. A private client will be more likely to throw a "topping-out party" after the joiner has fitted the roof truss, and then the later opening of the building. A public client will often launch the project, given that it is an important one, for the general public by turning the first sod, by arranging a "digger bite", or if the project is sufficiently large, by laying a foundation stone.

\\Note:
More detail on realizing a building can be found in *Basics Site Management* by Lars-Philip Rusch, Birkhäuser Publishers 2007.

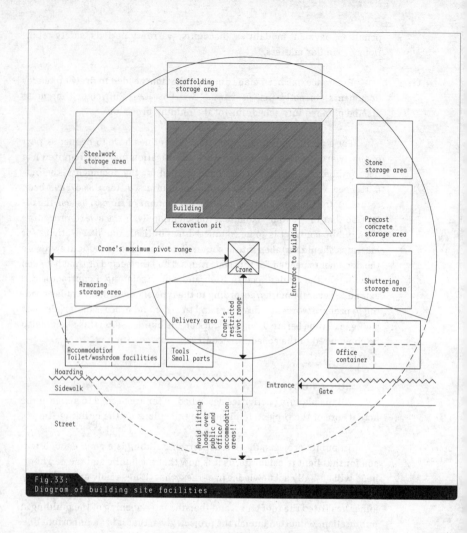

Fig.33:
Diagram of building site facilities

Diagram of building site facilities

The first measure to be taken is setting up and securing the building site. A scale site plan should be drawn up, providing information about the size of the area to be built on, the position of the site fence, the access gates, deliveries, storage facilities, working areas, scaffolding areas, crane positions and swing radius, locations for rest and rubbish containers, and the excavation pit. Connection points for building power and water must be established and also drawn in where appropriate. As a building site will make considerable impact on the surrounding area during the construction period it is essential to reach an agreement with the local authorities,

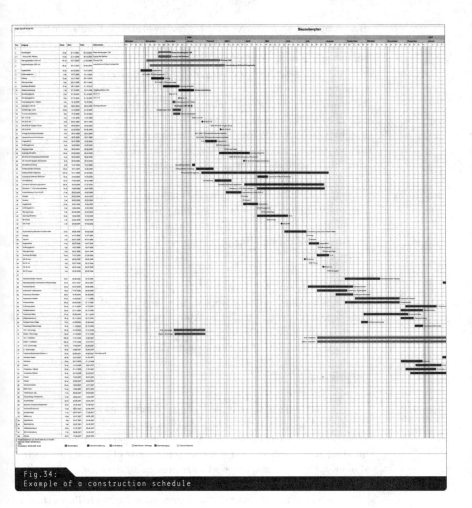

Fig. 34:
Example of a construction schedule

and contact with the immediate neighborhood is recommended. The authorities should be informed in good time about when building is actually due to start.

Scheduling

One of the most important elements of the site manager's job is to draw up a schedule, usually in the form of a bar chart. › Chapter General conditions and aims, Deadlines This coordinates the main building processes and trades, bearing the completion date in mind. It is important here that the individual trade jobs are brought together and effectively matched. If the first version of a working schedule is reviewed after the project is finished, it is likely that a number of deviations from the planned flow will be found. An enormous number of things affect the schedule, including periods of bad weather, awarding problems, delivery bottlenecks for materials, and possibly even insolvencies during the building period and delays caused by contractors. As the schedule is based on a logical sequence of jobs, delays by a single trade can often affect subsequent trades and the whole building process considerably. The site manager has to keep an eye on all components so that countermeasures can be taken in good time, and provide information about postponing the completion date.

Intervening in the schedule

If a schedule is to be properly maintained, it should allow for any necessary interventions into the building process. The client's stated completion date is unlikely to be met if there are building delays shortly before the end of the building period, so early interventions should be made where needed.

Interventions can take the form of raising worker capacity or increasing the length of time worked. Dividing the work into sections can also be helpful, so that trades working later in the sequence are not held up. It is also possible for part of the work to be done at a later date, at least if this does not affect intermediate deadlines or later work. As a last resort, the quality of finish can be modified in order to speed up the building process.

General site supervision

The presence of the site manager on site depends to a large extent on the size and complexity of the project. The architect will usually look after small domestic projects him-/herself, without a separate on-site office. But building projects on a larger scale need an on-site office and site supervision by one or more managers who are present all the time.

The site manager needs all the current plans and the building contract including the specifications for the jobs that have already been awarded for his or her dealings with the contractors. It is absolutely essential for the site managers to familiarize themselves with the plans in detail, so that the work can be supervised and directed according to requirements.

On-site meetings

For larger projects, the site office is also used as a place for meetings that take place regularly. The specialist planners and the firms working on the site should be invited, as needed and according to the way the work is progressing. The meetings should be minuted on the basis of jour-fixe minuting standards. ＞ Chapter Planning process, Working plans and quality of finish

It also makes sense for all new participants in the building process to exchange address data at an early stage, though care should be taken that the information is always passed on via the site manager responsible.

Construction diary

The site manager is obliged to keep a construction diary as a record of the work done on site. As the construction diary is often used subsequently to provide evidence, the following entries are of considerable importance:

- Date
- Weather conditions (temperature, time of day, etc.)
- Firms working at the site
- Number of workers per firm
- Nature of work done
- Orders, statements and acceptances
- Handing over plans, samples, etc.
- Material deliveries
- Special events (visits, hold-ups, accidents, etc.)

Presence

As has already been suggested, the presence of the site manager depends largely on the project in hand. For example, keeping a construction diary does not depend on the site manager being present every day. It has only to be brought up to date if the site manager is or has to be present on site. On-site supervision has to be concomitant with keeping to the permission requirements, the working plans and the job specifications, observing the recognized codes of practice.

There is no doubt that the site manager must be present at important and critical phases of the work, for example if sealing and insulation work

\\ Note:
Even the architect, if he or she is not one and the same person as the site manager, should not give instructions directly to workers without including the site managers. A building process is so complex that individuals involved cannot have a complete grasp of all the interlinked procedures and events. To avoid disagreements between the parties involved, all information must be conveyed via the site manager, who is responsible for overall coordination.

\\ Example:
If the responsible site manager is present on a prestigious building site twice a week, and can keep to deadlines and guarantee the quality of the work done, the client will have no grounds for objections. But errors of coordination can creep in even if he or she is permanently present, or building supervision could leave something to be desired. An architect conducting site management also owes "success" to his or her activities in this field.

69

is being done, reinforcement materials introduced, concrete of the required quality being delivered, for loadbearing constructions with appropriate anchorage, for excavations, if heat insulation is being fitted, when constructing sound insulation, etc.

But there is no need for site management to be present for simple, routine work. An experienced site manager can limit his or her presence on site appropriately.

Health and safety officer

The site manager is obliged to attend to traffic safety and hazard prevention on the building site. He or she is supported in this by the health and safety coordinator, who makes regular patrols to check that accident prevention rules are being followed, and to point out potential danger to the site manager if they are not. › Chapter Project participants, Experts

Specialist site management

The site manager has on a limited role, essential dealing with coordination, in terms of specialist firms working on the site, who are supervised by specialist engineers. He or she should on no account take on specialist management for trades operating beyond his or her competence.

Coordination

The site manager is the key link in coordinating work with the specialist trades. Two examples illustrate the importance of the "site management" control center. Only the site manager responsible can coordinate the interlined processes between placing dry partition walls and boarding one side of them, installing electrical equipment or additional plumbing and subsequent closure and painting of the walls, and then the final fitting of the electrical switches, as the individual trades have no cause to consider the work done by other specialists afterwards. Another classic case where more coordination is needed is parallel planning of underfloor heating with cable ducts under the screed. All the trades involved have to know about the timing for the work sequences, allowing time for the screed to dry and setting up heating protocols. This information has to pass via the responsible site manager, who then devises a sensible and feasible sequence of events, together with the specialist planners.

Acceptances

In the course of the building project, both the architect's work and that of the firms employed is appraised. By accepting the work, the client is stating that he or she acknowledges the service rendered. This process can have considerable legal consequences. As a rule the building is accepted after completion of the shell or at the final completion stage. Compliance with the permission documents or with the current building laws is tested at this point. Acceptance of connection with official supply grids is carried out with the suppliers responsible for water, sewerage, power, gas, etc. The heating system is accepted by the local master chimney sweep. Special technical facilities, for example conveyor systems or elevators need to be accepted separately by technical monitoring services.

Compliance claim and warranty stage

Formerly, the client had a "compliance claim" vis-à-vis the contractor or trades, and risk of loss used to lie with the contractor as well. "Risk of loss" means the risk that the work has to be done again without remuneration because of an event or damage that takes place before acceptance.

Appraisal

An appraisal of planning work can also be demanded, depending on different local legislation, and in a way that differs from building acceptance. Here the completed work is assessed for compliance with the plans as originally passed after completion and before the building is handed over and used.

The warranty stage begins at the point of acceptance. This will be dealt with in more detail in the next chapter. During the warranty or limitation period any deficiencies faults that occur must be made good, so long as this does not involve disproportionate effort or expense. The warranty period starts again once the deficiencies have been made good.

Inspection

Trades will not usually wait until the work is complete and accepted before invoicing the client. Hence "payments on account" or "part payments" are usual, according to building progress and the work done. Payments are usually made on the basis of the working plans. If a particular piece of work is not recorded there, or deviates from it, a joint inspection will take place on site with the worker so that the invoice submitted can be checked correctly.

Checking invoices

The contractor will formally address the actual invoice to the client, but will send it to the architect, who is responsible for checking it professionally and computationally. He or she may have to correct the invoice, and forward it to the client with a certificate that it has been checked. The architect must always remember when checking that he or she is the client's agent, and that thus it is not his or her duty to examine the invoice to the advantage of the contractor, in other words to the client's disadvantage. Ultimately the certificate on the invoice is just a recommendation that the client should pay the invoice.

Statement of costs

As invoices are now increasingly coming in from the trades, the client is obliged to make large sums of money available, as well as for the fee payments already due for planning the building project. He or she is thus increasingly interested in being aware of the current cost situation. The

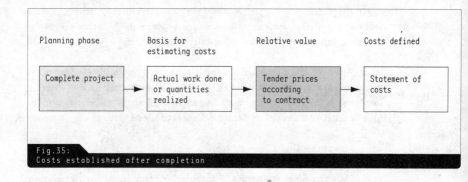

Fig.35:
Costs established after completion

budget estimated for the project is usually limited, dependent on credits, subsidies or permission from committees, and thus cannot be exceeded at will. Comparing the commission sums as related to offer sums and building contract with the actual sums invoiced after the various trades have completed their work is thus crucially important to the client. Furthermore, if costs are constantly monitored, it is possible to intervene effectively if costs are in fact exceeded. Thus the cost situation will come closer to the actual costs after the building is completed as building progresses. After completion the final costs can be summed up in a statement of costs. › Fig. 35

Handover After acceptance, as described above, by authorities and of the trades' services rendered, the property is handed over to the client. This does not have to be accompanied by a tour of inspection of the building. But the client does have the right to request that all the required planning documents are handed over. This documentation includes a full set of working plans, as well as installation plans for services, technical operation information, reinforcement plans, acceptance protocols and certificates. The documents will be more or less numerous according to the building, but they must be compiled in such a way that they can be used correctly. All the information for subsequent maintenance work, refurbishment or change of use must be available. Here the specialist planners are also required to make the necessary information available to the client.

The architect's duties are not absolutely completed with handover. There may be some work remaining to be done, and deficiencies complained about in the work done by the contractors have to be monitored and handled.

Opening Generally it is hoped that completion of a building is something that should be celebrated appropriately. The architect hands the building over

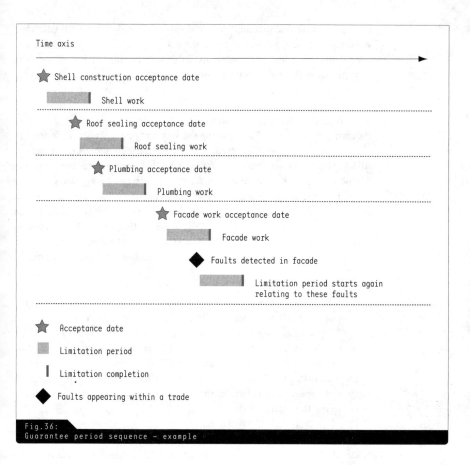

Fig.36:
Guarantee period sequence – example

to the client officially by presenting him or her with a symbolic key. In the case of pubic projects, this occasion can be used to attract attention and enhance esteem.

But apart from this, the commitment shown and services rendered by everyone involved in the building process can be acknowledged. The architect should also pay tribute to the client. The architect owes his or her fee not least to the client's readiness to invest, whether privately or publicly.

THE WARRANTY PERIOD

The architect can continue to perform valuable services for the client after the building is completed. A detailed and systematic documentation of the building, independent of the document handover described above, contains a structured record and analysis of all the data used in the planning. This does not mean coming up with new planning work of drawing up new plans, such as presentation plans, but merely brings together all the documents dealt with in the course of the project planning process.

Warranty periods

A crucial part of the work required after the building has been handed over consists of monitoring the warranty periods for work done by the individual trades. The services rendered were almost all completed at different periods, and thus also accepted at different times. The warranty period starts with acceptance. Different periods can be agreed according to the basis of the contract. Hence it is vital to put all the warranty periods needed for monitoring very carefully. Shortly before the warranty period is over the architect must make a tour of inspection to check the building for any deficiencies, and examine the structural components thoroughly. However, it is not necessary to use special examination methods or equipment.

Eliminating defects

If the examination reveals defects, they should be notified to the client. An appropriate time can be allowed in order to correct them.

Final examination

It can also be helpful for the architect to conduct a final examination of the project with reference to cost guidelines, related to square meters, cubic meters or individual trades. It can also be useful for planning future projects to examine the costs incurred and the time needed for the practice to complete the work.

The work needed during the warranty period can take a disproportionately large amount of time in relation to the fee due. But this time and effort can be limited effectively, not least with aid of good planning and the right selection of competent building firms and workers, as well as through conscientiously discharged site management.

IN CONCLUSION

When planning the project, the architect takes responsibility for realizing it successfully. The project manager responsible must have his or her eye on the initial aims throughout every phase of the project. Planning requirements in terms of cost, schedule and quality are the coordinates at which work must always be directed.

The architect is the client's expert contact person in all fields of building. He or she will advise the client about building planning questions, and the relevant technical, economic, creative, civic, and ecological aspects. He or she functions as the recognized and qualified coordinator between everyone involved in the building project, such as specialist planners, contractors, workmen, local authorities and departments.

The architect introduces the planning steps in logical sequence and arrives at the decisions needed at every planning phase with the client.

In the realization phase he or she checks the sequence of events to ensure that all the aims formulated at the planning stage are achieved.

At a time of increasing specialization in every field, realizing a building project requires a wide range of knowledge and ability from the planner. Being an architect is an attractive profession not just because of the creative design possibilities, but largely also because there are so many different challenges. As the previous chapters explain, financial and strategic skills are extremely important. And architects must also address all the relevant legal matters in the course of their work. The ability to work in a team and to be able to deal with all kinds of people requires a high level of social competence. The architect will come into contact with a large number of professional partners in the course of the project, whether it is the partners involved or the trades who will work on implementing the building project. Each building project means coming to terms with the implications of new and interesting fields of work. And not least a planner needs a marked ability to articulate his or her designs and ideas responsibly to society, which he/she is contributing to designing.

Realizing a project from the idea to handing the building over embraces almost all of an architect's field of activity. Every new planning stage brings new challenges. Successful completion and an effective opening confirm and underline the planner's achievement.

APPENDIX

PICTURE CREDITS

Figures 10, 17, 20, 29	Bert Bielefeld, Thomas Feuerabend
Figures 22, 24, 26, 28	Tim Brandt, Sebastian Th. Franssen
Figures 23, 25, 30	Udo Blecken, Bert Bielefeld
Figure 33 (Diagram of building site facilities)	Lars-Phillip Rusch
All other figures	The author

THE AUTHOR

Hartmut Klein, Dipl.-Ing., practising architect in Freiburg im Breisgau specializing in managing public building projects.

简介

　　一个工程项目起始于将一种三维构思、一种空间或一项地产投资的需求转化为建筑实践的意图。此类"项目"的提出，既要有对令人信服的高品质创意的渴求，又要有坚持设计概念直到工程完工的意愿。而项目规划的目的就是完成设计意图，将构思变成现实。

业主/建筑师

　　正如每个项目都来自委托人，建筑亦不例外。业主委托某位建筑师拟定一个设计方案，用以筹备、规划、监管以及实施建筑工程。一个典型建筑师的核心任务是在从调查基础工作开始，完成设计到规划、投标、现场管理以及建筑完工等项目规划的一系列工作。对于业主和建筑师来说，建筑成本、完工日期和完工质量都是息息相关的。

构思/实现

　　建筑师在规划一个工程项目时所需的初始动力会以多种方式出现。在许多案例中，建筑业主或投资人会把他（她）的所知直接告诉建筑师，用或多或少的细节来说明他们对项目的想法和要求。工程项目的获取通常是采取竞赛或专家报告的形式，一些建筑师的设计参与竞标，由业主基于之前的投标要求作出选择。

　　但反之亦可以想像：建筑师联系潜在的业主，在一个可接受的条件下让业主作出委托可能。这种情况需要建筑师细致的研究，以便找到那些正在或者是不久的将来需要建造房屋的业主。

规划步骤/决定层面

　　为了保证项目从创意构思到实际完工和交付使用的进度，在细节日益增长的复杂情况下必须逐步计划好并确保实施。最初抽象的概念在各个阶段要得到逐步充分的阐明、具体化和贯彻执行。

　　方案最初的框架绘制于纸上。随着相关人员不断增加，草图变成了一定比例的图纸平面，平面成为实际施工的基础。经有关当局颁布工程许可后，建筑公司和建筑商受邀竞标，最后，承包人拿到工程委托。这是开始实施建筑项目的第一步。一旦不同的行业在建筑工地上合作成功，盖起一个房子的目标就会成为现实。

　　如果一个建筑项目要规划和执行得明智且有远见，需要采取诸多步骤。根据项目及其结构和体量之特殊性，这些步骤也会有所区别。但是每个项目案例中，即使责任分配的方式多种多样，事件的总体进程却是相似乃至具有普遍性的。因此，在德语国家，建筑师在建筑设计到建筑完工并移交业主的整个过程中，始终扮演着承担核心责任的角色。然而在北美和其他许多欧洲国家，责任方在设计阶段过后即移

决定层面

交到其他合伙人身上了。

根据不同的项目阶段要求，业主必须在多个层面作出决定。

1. **项目实施决定**：为了从根本点着手一个工程项目，多个参数（例如建设用地、建筑功能、资金和周期）有待检验是否符合他们的基本承受力。项目的实施决定由项目计划相关的参与者作出。

2. **构思采纳决定**：如果建筑师最初的构思（包括功能分区、体量和面积说明、毛成本等参数的框架）已获认可，那么业主必须决定是否要求建筑师进一步深化最初的设计理念以满足他（她）的需求。

3. **工程申请递交决定**：一旦设计构思包括以上提到的参数已被深化，业主必须决定是否向建设主管部门递交现有设计规划的申请，因为申请如果获准，现有设计在日后修改的幅度将很有限。

4. **施工质量决定**：项目获得建筑许可后，即进入工程筹备阶段。在这种情况下，业主必须在工程质量和材质外观之间作出决定。这个决定通常是基于材料测试、过往案例、材料说明以及造价调整说明等因素综合考量后而作出的。

5. **工程合同授予决定**：在之前的各种决定已经作出和承包人实施提案汇总的基础之上，准备项目施工文件。然后，业主必须决定哪个承包人可以获得相关工作合同，抉择的参照则是建筑师的评估和推荐。

项目参与者

毫无疑问，工程项目中最重要的两个角色是拥有委托权的业主和制定项目规划的建筑师。但建筑的规模和建设目标决定了项目进程中还会有大量相关的合作者参与其中。>见图1

业主

业主/使用者

业主是拥有建筑筹划和建设权的个人或组织团体。从法律意义上来说，业主是民法或公法下的自然人或法人。

业主和使用者可能是同一个人，这取决于即将竣工的工程属性。如确为同一人，则建筑师应仅从一个人那里寻求项目意图并征得同意。然而，尤其在公共建筑项目中，有时可能是私人开发商，建筑师经常遇到业主双方意见相左。此外，还有更深一层的决定层面，如注入外部资金的投资方或者业主委托的管理委员会，都能够影响到规划

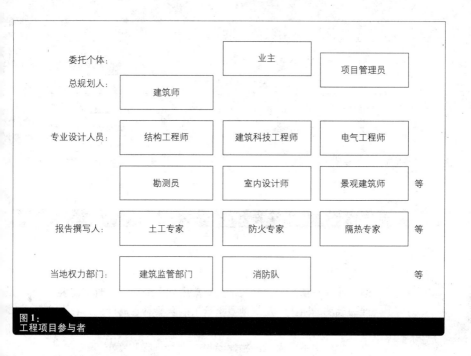

图1：
工程项目参与者

和建设的进程。随后的使用者也取决于工程性质，他们可能是学校的老师，可能是消防站的消防员，也可能是医院里的医生或护士。业主通常会邀请这些未来的使用者们参与到工程前期的规划阶段中来，但有时使用者们的要求会超出业主的责任范围。所以，一个工程项目要想取得成功，很重要的一点就是要在设计构思和实际需求两者之间找到平衡，作出合理的判断。 > 见图2

建筑师

在建筑工地，一般是建筑师或专业设计公司提供设计咨询服务。

建筑师是回答所有建筑问题最合适的联络人，他（她）为业主在项目实施过程中有待解决的所有事务提供建议。作为业主的代表或代理人，建筑师会与工程进程中所涉及的每个人打交道，包括权力部门的人、承担部分专业设计方面的专家或公司人士以及商贸人员。

建筑师严格审视业主需求的合理性，根据项目的财政状况提供建议和构思，并附以一个合理的日程表预估和可能的设计变化，和业主

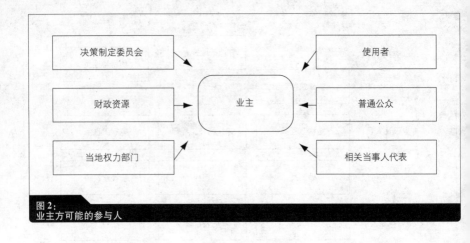

图2：
业主方可能的参与人

一起逐步探讨，寻找解决办法。建筑师的部分工作就是拿出令人信服的设计并成功实现。当大量人力集中到一个项目上时，需要一个高水准的成功的团队。建筑师要求这个团队要具备高标准的社会竞争力和高素质的人员业务能力，进而领导和指导他们去实现项目。每个规划设计阶段中不同需求的描述参见"规划进程"章节。

P13

项目管理员

项目中有关专家人员的数目随着工程大小的不同也会有所增减。如果一个项目的规模和周期超出了业主能力和专家意见的范围，那么最好增加一名项目管理员。

项目管理员拥有业主在技术、经济和法律方面可以委任的职能。换句话说，管理员就是业主的顾问，但通常没有权力代表业主的法人地位。＞见图3

项目管理

项目管理员的工作一般不会涉及建筑师本身的规划设计任务（如初步设计、设计规划等），但与整个工程的管理工作却是息息相关的，从财政分析开始，包括资源提供、合同处理以及员工管理等一系列工作。

对建筑师的支援

在一些大型项目中，项目管理员可以在工程管理方面给予建筑师有力的支援，并帮助协调和管理工程相关的人员。

P14

专业设计人员

在相对小一些的项目如独立式住宅项目中，建筑师几乎可以完成

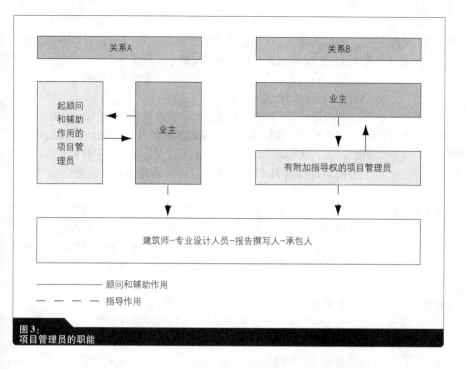

图3：
项目管理员的职能

工程设计所需的所有技术工作。但同时，如果要保障设计实施的顺利，两个合伙人也是必不可少的。

勘测员　　一个是特许勘测员，通常受委托画出正式的项目场地平面图。场地平面图不但是建筑部门批准项目程序中的一项要求，还为日后建筑测量所需。

结构工程师　　另一个合作者是结构工程师，合适的有资质的结构工程师负责计算建筑中相对稳定的部分——楼面板、墙壁、顶棚和屋顶的设计尺寸。

在小体量建筑像独立式住宅项目中，建筑师经常也会承担电气装置、暖气装置、卫生设施以及室外区域等部分的规划设计工作，或者是与受委托的公司和商人们共同完成。

当遇到大规模建筑项目如公共建筑（体育馆、市政厅、消防站等），大型办公建筑群或著名公司的建筑时，所有设备的规划设计工作均需由各专业的工程师完成。

建筑设施　　工程师的工作任务还包括设计供热技术和卫生系统，即做好给水和污水处理。同时还承担通风、冷却或空调装置、煤气设备

	的设计。
电气工程师	电气工程师不仅仅为建筑配置电路和照明，还同时负责做好防雷电设备、防火防烟警报以及标识清楚逃生路线等工作。
	对于一些著名的规模庞大的建筑，建筑师工作中的很多部分通常会转交到由不同领域的专业人士来完成。
室内设计师	室内设计师受雇设计特殊区域的风格或布置每个装饰构件。
景观建筑师	园林和景观建筑师在与建筑师协调的情况下设计和规划外部空间。这里，景观建筑师的工作对象不仅仅局限于奢华的私人花园，还延伸到功能性公共广场或绿色空间、体育场地，甚至是噪声防护设施。
立面设计师	在一些大规模的群体建筑中，建筑师会建议业主聘用一位专业的立面设计师，以避免多个单体建筑样式纷杂造成的立面冲突。
照明设计师	照明设计师或照明规划师受委托设计特殊的和普通的照明。不论昼夜，他们模仿和制造着技巧性和创造性的灯光效果。
P15	专家组
	专家组，不像专业设计人员，不会提供具体的规划设计方案。他们的角色是顾问，针对方案中出现的问题，提供各种报告用来描述问题的情况或确定问题发生的原因，并提出解决方案的建议。几乎任何领域都可以吸收进专家组的意见。以下就是建筑活动中最重要的一些领域：
土工专家	根据当地土质状况和天然地基情况，需要土工专家或地质学家依据试挖掘和试钻探或现存的地图，提供基础层施工的可能性和现存地下水的资料信息。

> 注释：
> 在项目初期，建立一个有针对性的项目框架是十分有益的。所有专业规划设计所需的参与者都必须在合适的时机进入项目中。项目组织过程中的重要事项包括建立一个涵盖所有参与者联络地址的名单，确定定期讨论日期（jour fixe），起草限期内已完工项目信息的书面记录，以及不同参与者之间数据（DXF、DWG、PDF等数据格式）交换所商定的协议。

建筑历史学家	如果一个正在重新整修的建筑自身具有历史价值，那么聘请一位建筑历史学家作为顾问则是很有必要的。历史学家会把该建筑的历史沿革汇编集成，并提供建筑结构中值得保留部分的评估报告。这样做至少可以确定修复范围，也就是确定重新整修所限定的历史时段。
交通规划师	如果一处建筑工程影响到当地的交通状况，或需要改变现存的基础设施和交通道路，这时候就需要交通规划师的介入。
防火专家	在一些建筑工程，特别是规模大需求多的工程项目中，做规划时有必要聘请防火专家。他们可以提供符合法律规范的至关重要的规划信息，同时，为获取工程许可必需的资格，还要拟定防火报告或设想，并监督这些报告或设想是否被正确的执行。
隔热、隔声专家	对于许多建筑类型来说，委托隔热隔声专家是很有意义的。专家们不但能为新盖的建筑解决隔热、防潮、隔声等问题，还可以评估现存建筑中的缺陷和残损。
声学家	声学是建筑物理性能评估的另一个方面。在一些声学标准要求高的场所，如阶梯教室或音乐厅，需要计算出最好的声学效果，而处理冲击声、空气载声和脚步声隔离问题则相对简单一些。所以，在建筑设计阶段，建筑师要与声学家保持必要的联系。
防污染专家	来自防污染专家的建议尤其适用于现存建筑，也就是为建筑的保护、修整和改造而提供的有价值的建议。他们可以检测和评定出建筑已经使用的构造材料。目前的研究结果表明，那些影响建筑使用者和居住者健康的材料日常还在使用。而在建筑重新整修和去除有害建筑材料时，也会导致一些对健康造成影响的特殊问题。 防污染专家的研究和建议对于正确的投标和有害材料的安全处理都是十分重要的。
健康和安全协调员	欧盟（EU）建筑规范强调一旦建筑工地超出既定范围，就必须聘用健康和安全协调员。这项措施可以由业主或建筑师执行，只要他们有足够的资质，但也可以由另一个单独的人来执行。这就意味着建筑工程在设计和施工过程中在安全方面将会被监控，从而保护业主和承包人的员工以及无分红权的第三方尽可能远离危险。
P17	**相关部门和权力机构** 迄今为止，以上提到的专家仅仅是在项目规划设计阶段所涉及的。那些签发工程所需各种许可的机构和权力部门在不同的阶段还必将会提及，因为没有他们，项目是不可能实现的。示例来说，即使在项目初期，市属或当地权力部门的介入也是有意义的，使用由他们提

建筑管理 部门	供的现有发展计划来规划项目总体发展，可以确保建筑更好地建成。 本质上，建筑师在项目进行中也是要与建筑权力部门打交道。他们在建筑可控程序内（规划许可）负责确保建筑活动符合其他相关规范。 根据建筑的规模和建设许可需求，其他部门也可能在项目过程中起到作用。这些部门包括地政局、测量机构、土地储备办公室、市镇规划部、遗址保护局、环境部、土木工程部、城市园林局等其他很多部门。对于公共建筑来说，项目的设计通常要遵守当地消防局或防火防灾办公室的规定，这些规定包括相关规范信息以及火情和援救的要求。（见"规划进程 – 获取许可阶段"章节）
P17	**承包人** 承包人，理所当然是建筑项目中业主重要的合伙人，同时还是设计者和相关负责人。专业人士和建筑公司根据可行性方案和方案实施说明在现场操作施工。本质上说，以下两个不同模式都适合委托业主：
单一服务	第一种模式中，针对每个单一行业提供的专业服务，业主可以将项目的部分内容单独授予能够提供有关服务的公司。（行业服务的意思是由某个行业工艺部门提供的专项工作）
总承包人	另一种模式中，业主可以将整个项目合同完全交给一个独立的承包人，即总承包人。总承包人在所提供项目的基础上亲自执行或是雇用分包人运作实施工程项目。（见"规划进程 – 投标、施工"章节）
P18	**合同框架** 业主确定规划设计和实施工作的委托合同具有多种可能性框架。 > 见图 4
规划阶段 合同	—业主与每一个项目相关的设计人员和专家单独签署合同，然后由建筑师与之合作； —业主与能够提出设计规划的建筑师签署合同。随后，与投标和现场管理有关的规划要求转交至专家，并应用到实际规划中，同时，专家又将自己的意见和适当的专业规划设计提供给业主； —业主与总设计人签署合同，总设计人的任务是委托其他所有相关专业的设计人员和专家作为项目分包人。因此，业主在整个工程项目进展过程中只需要与一个合同伙伴单方面联系即可。在这种关系中，项目分包人承担的责任是代表另一个承包人（总承包人）为业主完成其所涉及的相关专业工作。

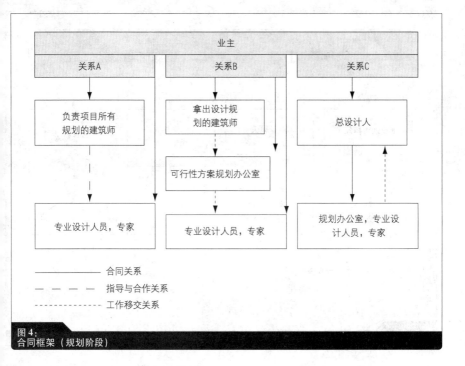

图4：
合同框架（规划阶段）

实施阶段
合同

实现委托项目的多种操作方式：>见图5

—总规和详规完成之后，业主与所有相关专业人士单独签署合同，建筑师再与每个行业人员进一步合作；

—总规和详规完成之后，业主与提供项目所有服务或雇用分包人的总承包人签署合同；

—递交设计之后，业主与总承包人签署合同，总承包人完成项目所需的技术服务（最终规划、结构设计、专家规划以及专家意见等）以及建筑施工，或者聘用分包人实施；

—业主与总承包人签署合同，总承包人承担的责任包括项目所有的规划，具体的设计及随后的施工，以及聘用所有与项目有关的参与者（包括建筑师、设计人员、专家和施工公司等）作为分包人。

汇总以上描述的委托方式，根据国情的不同，具体操作起来也会有所调整。委托规划和运作实施之间界限的阐述也会有地域差别。但是对绝大部分综合项目而言，规划步骤的顺序是相似的，因为项目在规划阶段本质上是要在合同伙伴自身属性之外独立执行的。

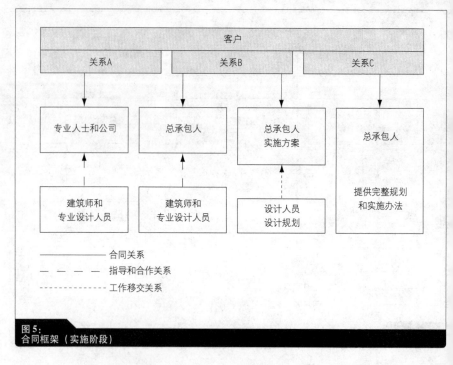

图5：
合同框架（实施阶段）

团队建设

从业主和建筑师之间的第一份合同开始直至建筑项目完工，所有与项目相关的事物在不同的阶段都要经历缜密的处理过程。从一开始就奠定合作基础是十分重要的，因为这样可以在项目进程中更好地处理随时可能出现的棘手问题。在实施项目的过程中，在得到认可的经济成本参数和工期日程内，业主获取利益是无可非议的事，追求利益并不意味着只是为了表面的和谐协作，而是意味着要让在一起从事富于建设性工作的参与者们积极干好手头的工作，并让合作者们互相感到满意。建筑工程从来不是地下活动，所以，参与项目的设计人员和权威专家，甚至业主和投资人，均表达出一种普遍意愿，这种意愿也体现在所有社会协作的精神中。>见图6

规划团队

项目的规划设计和施工可能要持续几年时间，这取决于项目的规模，如果条件允许的话，可以分成几个建筑周期陆续完成。所以，项目规划小组及其潜在的理念和基本构想应该能够灵活应对各种各样的情况和挑战。规划一个项目要有周密细节的考虑和卓越的远见，但随着项目的进展，新的见解和挑战也会随之而来，同样需要相应的对

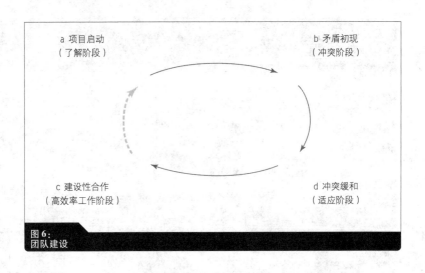

图6:
团队建设

策。项目进行中,会有越来越多的设计人士加入,有关权威人士和部门也必不可少,以及不断增加的公司和各种技术人员,他们作为投标和项目委托的机构或个人在项目实施期间发挥着各自的作用。

团队能力

每个工程中不时冒出的角色(业主、建筑师、专业设计人员和技术人员等)通常是由变化着的合伙人和个体扮演的。建筑师作为协调人,在建筑团队进程中起着重要的作用。项目团队,像其他每个工作团队一样,经历着各式各样的社会行为阶段。

最初的阶段,人们经常会像预期设想的那样彼此客气,团队成员们易于兴奋、好奇并期待互相更好的了解。但是手头的项目要求每一个人做出影响别人的成绩来,这种情况容易导致工作和个人之间的冲突。如果大家的工作关系是对峙和紧张的情况,而人们又绝不能失去他们的客观性,那么这时,作为中间人和协调人的建筑师,就变得十分重要了。在这段所谓"适应阶段"中,有必要让大家互相理解每一个人都是在朝同一个目标努力工作着的,而且这个目标只能靠大家把力量拧成一股绳,在交往和交流中以互相尊重的方式去实现。同时,对专家意见的质疑精神绝对不能失去。

最理想的工作基础一旦确立,团队成员应该以信任、坦诚、独创和团结的精神互相交流合作。这样可以使项目小组更有效、更有动力、更有针对性地实现项目目标。所以,建筑师工作任务中必不可少的一部分是要给项目小组清晰的定位,并辅以高效率的项目管理。如果没有这个定位,项目小组可能会迷失前进的方向。

概况和目标

设计人员在每一个项目的规划阶段不得不处理三个基本目标。实现这些目标对业主和建筑师都是至关重要的。首先，第一个目标是业主界定的成本框架和建筑完工期限。业主乐于通过建筑品质来传达他们对待建筑的态度，而建筑的品质由两个参数直接决定，即<u>预算</u>和<u>工期</u>。

项目成本

建筑的成本对于业主而言是非常重要的，所以他们希望建筑师从项目一开始就能够在造价控制方面给他们提供合理而又专业的建议。

业主对成本保障的推断源自建筑项目所需达到的目标。商业、公共和私人建筑业主在阐述他们各自目标的时候是不同的。

<u>商业建筑业主</u>在制定成本框架时，追求的是经济目的。对投资的业主或第三方来说，预期的成本效益是最重要的考虑因素。另外，利润回报率和未来保值率对于这项投资也很重要。

<u>私人建筑业主</u>是投资他们自己的未来，他们花的是自己的钱。所以，已规划项目的未来发展潜力是很重要的考虑因素。而业主的资金来源和稳定的投资实力在项目实施时也应该给予考虑。

<u>公共建筑业主</u>的投资意图来自提供基础设施建设的责任，基础设施反之也服务于他们的投资项目。根据管理范畴和建筑结构的不同，投资性质也有所区别。示例来说，来自地区或当地社会的建筑委托可以与安全紧急设施、防火防灾设施、健康保障设施（医院）以及教育设施（学校）等有关。更小一些的社区负责提供幼儿园、成人教育

> **提示：**
> 投资收益的细节阐述参见《主题：建筑成本和工期规划》（Thema: Baukosten – und Terminplanung），贝尔特·比勒费尔德、托马斯·福伊尔阿本德著，Birkhäuser 出版，2007 年。

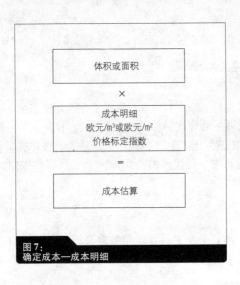

图7：
确定成本—成本明细

学院、社区中心和其他教育设施。公共建筑业主的投资决定往往是建立在需求评估以及社会目标或理想主义目标基础之上。决策措施的成本效益由最长的可能使用周期定义。但即使对公共建筑业主而言，这种措施的有效来源也是受到限制的，应得到对应的行政管理委员会的批准。

成本框架　　业主为建筑师提供成本估算作为项目实施的一个成本框架。这个框架可以被认为是受约束的上限，也可以被认为是大致的目标。建筑师基于此框架计算出合理的工程造价，并根据预算提供给业主关于房屋面积、体积和大致标准规格方面的资料信息。同时，自然界无法估量和无法确定的因素以及成本范畴都应该在资料中有所反映。

另一个可能需要业主为建筑师提供的资料是关于项目的功能、空间布局和预期质量的需求。然后，建筑师基于这些需求评估出项目的预期成本。

事实上，以上两个步骤经常是交叉进行的，因为业主会经常从其他领域和项目中核实成本，甚至在项目初期，在成本框架内维持项目的进行也变得日益重要起来。

成本明细　　建筑师通过多种办法对实际规划阶段的各项成本单独预算，项目所有成本的确定是数量因素和成本明细的乘积。成本明细反映了参考单位（如面积和体积）的成本比率，例如一个典型成本明细的规定是每 $1m^2$ 毛面积的造价是 1000 欧元（1000 欧元/m^2）。> 见图 7

知晓目前造价发展的水平，有助于业主在规划阶段作出决定。第

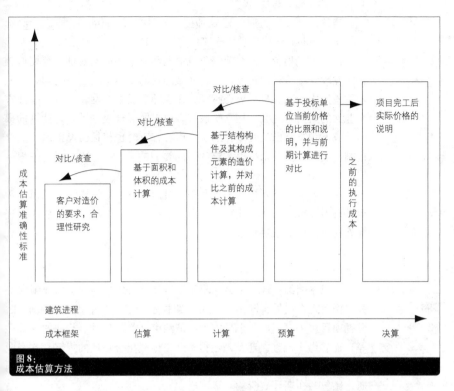

图8：
成本估算方法

| 确定成本的步骤 | 一步是项目起点的决策，判断设计构思是否切实可行，然后初步勾勒草案，在十分粗略的基准点上评估出一个可能的成本。这些成本估算的基准点包括所需的房屋使用空间、房屋毛面积或建筑的毛体积。如果项目规划要进一步推进，设计要有所变更，那么项目成本在决定提交建筑申请之前就必须计算出来。＞见图8 |

更准确的成本计算基于掌握承重结构和独立结构的细节资料，以及粗糙或非常精细的结构元素的基础数据上得出的。

结构粗糙元素可能包括内外墙体、顶棚或屋顶，精细结构元素是指粗糙元素内的每个细节元素（顶棚抹灰层、钢筋混凝土顶棚、找平层、楼面料等）。这一阶段确定下来的综合造价计算要比之前的成本计算更准确。

当承包人和投标人接到工程任务时，下一步就要提交目前的造价说明。造价说明要与之前委托的各行业单独预算相一致，目的是将真实的市场价格与成本计划相互比较，然后在有必要调整预算的地方进行干预。

当建筑项目完工时，建筑师汇总整体施工造价，这种<u>造价测定</u>（cost determination）记录了建筑的实际成本。

成本控制

在成本计划中，针对前一阶段情况和业主的成本明细，定期<u>检查新的发现和细节</u>是很重要的，这样做的目的是尽量在工作初期就发现矛盾差异之处。项目进展得越深，建筑师在成本开发和矛盾可控性方面遇到的阻力就越大。施工阶段有必要采取混合成本计算，计算的根据是成本预算和已接受委托的工作内容，有时也包括已经接纳并支付报酬但实际可能并未得以实施的不同类型的工种。（见"规划进程 - 施工"章节）

成本追踪

建筑师通过不断提供成本利用的进展信息，对最初预算谨慎尽责地进行追踪审核，从而获得业主们的信任，并告诉业主他们能够处理好委托给他们的项目资金。

P27

业主规定的期限明细

项目期限

与成本明细一样，项目期限明细也是项目成本的一个重要组成部分。项目现有租赁合同中，因为长期告示所需，可能做出期限通知，或者项目接手日期可能也会因为经济约束（产品启动、行业的季节性等）而要做出调整。建筑师不得不严格地提出合理性的规定，并经业主的同意，计划好时间期限的次序。这里需要考虑的不仅是建设周期，还包括初期计划阶段。

规定时间期限的基础通常是完工的<u>最终期限</u>。同样，也有可能在<u>建筑开工</u>时就详细说明一个时间期限，例如与资金补偿挂钩时。时间规定的另一个变量是<u>最短可能完工期</u>。这时，项目的开始和结束不必严格地界定说明。<u>短限完工期</u>应该尽可能地减少由项目建设周期所带来的约束和阻碍。

日程表范围

<u>项目日程表</u>的拟定是为了描述项目完成所需的完整时间，这需要考虑到计划周期和实施周期。项目日程表也有助于在建筑周期内出于流程清晰的考虑编制独立的<u>产品日程表</u>（工作日程表）。同时，建筑周期取决于规划时间和委托承包人时间的事实也必须

> **重点：**
> 业主有权要求获悉整个项目进展期间的即时成本报表。正如建筑师从一开始就关心规定好的成本明细，他们总是把成本列表展示得明晰易懂，以便与之前的成本信息进行对比。

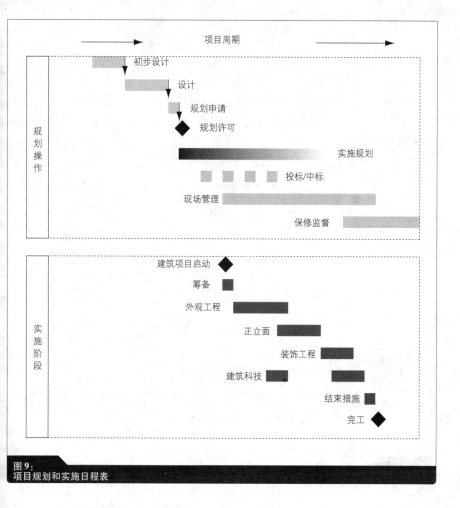

图9:
项目规划和实施日程表

规划计划

反复斟酌。>见图9

项目开始阶段,业主和建筑师会用大量的时间来理清思路,在项目协议要达成一致的复杂进程中还会牵扯到专业设计人员和相关责任人。因此,建筑师不得不创立一个工作基础平台,并将这个平台提供给所有相关的设计人员以便他们开展工作。反之,建筑师依靠设计人员的工作进展及其提交的完工记录文档作为项目的收益。
>见图10

表达方式

今天,日程表已由计算机专用程序设计完成,其最简单的表达方式就是一个由项目流程和起止日期组成的列表。但是在项目所涉及的

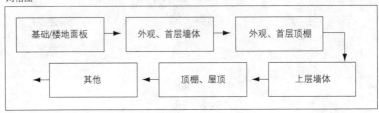

图 10：
日程表规划表现方法

网格图		流程大量涌现时，这个简单的列表就不是特别的清晰了。
		网格图说明了各种流程之间彼此关联和互相依赖的关系。这种方式很好地图解了流程间的串联关系，但却不能说明时间上的次序。
柱状图		最普遍的表达模式是柱状图表或甘特（Gantt）图表，Y 轴列举每个事件或项目流程，X 轴为时间轴。时间轴上的柱状图表示每个流程的期限。重叠和相关的部分毫不费力就可以得出，特别是通过现代的日程表程序。重大的连续关系可以由流程 A 的起始点至流程 B 的结束点链接表示，诸如此类进行。

96

转折点	转折点是一个十分重要的阶段或者说是整个项目日程表中（施工开始、结束等内容）很直接的一种期限规定。例如，施工许可就是日程表上一个重要的转折点。关于权威部门的记录检查，必须考虑留有足够的时间保障，因为施工许可上已经界定好的内容如果出现必要性改变的情况，工程开工就会受到影响。施工的转折性事件、表现结果和进程可通过柱状图表的方式，用色彩区分得更明确，对于所有相关人员更易理解。
目标执行参照	目标执行参照可以用来检查实际施工进程和日程表进度的关系，并确定项目是否依据原计划在进行。制作日程表的目的是在项目早期就判断最终的时间期限是否存在风险，并在适当的时候给予纠正。只有通过不断的检查和监控，才有可能及时地将正在进行的错误导向拉回正轨。
P30 技术标准	**项目质量** 原则上，质量的定义分两方面来看。总体来说，时下的标准、方针、规章和法律是项目在规划和实施过程中不可改变的基石。这些<u>实践得出的规范</u>已被业内专家普遍认可。基本上，作为技术标准，这些规范适用于所有的建筑项目并且无需在合同中特别阐述。无论如何，有效而又综合地应用好所有的标准和方针是项目品质得以保障的一个要点，在实施阶段也不能将规范视为理所当然的摆设，而是需要确确实实地贯彻执行。> 见图 11
独特的修饰标准	业主从个人角度列举建筑的各项标准，他（她）定义建筑的视觉外观，从材料的质量、形式和色彩的角度，在技术可能性的范畴内描述所有的结构构件。这时，业主的投资态度将是最终质量标准的决定因素之一，因为不同的完工质量，其投资成本的差异会非常大。相比之建筑材料的差异，建筑的色彩主题和形式差异对预算的影响要小得多。例如，立面或室内的表皮大面积使用玻璃要比使用密实砌体或钢筋混凝土更费钱。使用橡木作为窗框比塑料明显更显档次，但造价也更高。再如对比织布地面，有些特殊地面会安置高质量的镶木地板。这些例子屡见于整个结构构件的购置清单中。对于建筑师来说，很重要的一点是在工程项目初期就要和业主探讨完工标准，通过实例帮助业主作出决定，理清所需成本。> 见图 12
质量定义	就工程质量而言，已被认可的实践规范、方针和标准无需进一步地界定清晰，但每个完工标准需要以书面的方式清晰地定义，然而，项目初期的这个非装订书面标准是可调整的，例如"高标准"或

图 11：
国际标准化组织

图 12：
确立每个使用样本的修饰标准

"普通标准"，给日后留下很大的洽商空间。类似的标准说明在规划阶段将会不断地推敲完善。

首先，承重结构和必要的建筑构件的完成标准必须确定。随着项目的进展，详细的决策必须贯穿于每个构件的安置过程中。建筑师通过提供样本和比照成本内的各种完工情况帮助业主决策。一旦作出决定，执行过程的记录便根据一个总体定性的建筑方针或一个详细的房屋汇编簿而得出的，房屋簿里功能相同的空间可以被整合记录。数据单的内容包括房屋的命名、数量和楼层、墙体信息、顶棚、楼面、门、窗、暖气片、卫生和电气设施。

这种质量定义构成了成本计算的基础以及后期需要起草的行业服务明细的确定要点。如果业主在项目进程中欲增加或改变项目内容，建筑师必须指出相应变化带给成本方面的影响，参照对象就是业已界定的完工标准。

监管质量　　项目进程中，现存的技术标准和每个确定的完工标准均由工地现场经理监管。同时，质量控制意味着需要检查建筑材料的交接和安装、监管技术质量、按照指示操作制定细节以及遵守所有现行规范。正常的尺寸公差方面的考量也是整个质量监管过程的一部分。

一栋新建筑的品质不仅反映在令人信服的设计上，还与技术标准的正确运用和对业主个人需求的成功执行密不可分。

P33

规划进程

P33

项目决策

在一个工程项目的开始阶段,建筑业主必须解决好大量的基础性问题。首先,项目的财政状况必须接收审查。另外,如果这个新建筑的建设地点不能固定,那么一个合适的土地地段必须确定。业主还要确定工程项目的完工日期及所需的成本框架。在过去,这些问题主要是由业主自己在实际规划阶段开始之前予以解决。随着建筑业主工作内容的总体情况愈发复杂,即使在初期规划阶段,建筑师现在也已经成为一个重要的联络人了。> 见图 13

工期和预算

一个项目所涵盖的工程内容在初始阶段就必须阐述清晰,业主的想法和意图也需要鉴别和审视其合理性。这种审视既包括对业主经济实力的审核,还包括项目完工所需时限的合理评估。所以,示例来说,建筑师不得不参照实际案例评判业主的想法最终能否实现,并和业主探讨一个现实而又适当的实施方案。

建设用地商议

建筑师可以就一块地皮的选择或有效地段的评估给业主提供建议,就有关规划建设项目地段的位置、特点、环境和其他情况提供注意事项和建议。但是参与获取地皮或为财政变化出谋划策通常不是建筑师分内必做之事。

权威专家和专业设计人员

一个没有经验的业主应该被告知与项目有必然关系的权威专家和专业设计人员的情况。他们必须知道实施建筑项目需要增补哪些相关的设计人员,工程开工前需要获取或完成哪些报告、分析和调查研究,因为在大部分情况下,业主还需要与以上相关责任人签署合同。

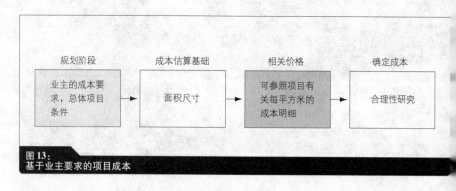

图 13:
基于业主要求的项目成本

（见"项目参与者－合同框架"章节）

参与者相关设计工作范畴

不是所有的业主都熟悉一个建筑项目进程中的每个步骤和发展阶段，但建筑师能够帮助他们的业主作出决定，尤其是当选择专业设计人员如结构工程师和建筑技师时，建筑师会根据专业人士的能力以及他们所能提供专业服务的范畴和所需资金，为业主提供建议。当然，建筑师自己的职责也必须被界定清楚。此类合作协议应该尽早签署，以便委托和承担的工作过程中相左的观点在后期不会出现。

获得项目和委任工作的任务

在一个工程项目的初期，建筑师的角色定位是获得业主项目和委任他人工作的中间人，所以他（她）应该尽最大可能配合业主实现项目需求。这个探索阶段通常被视为没有报酬的委托任务阶段，也就是说在一开始就应该把基础情况给潜在的业主说明清楚，诸如成本、日程、地段以及相关人员的选择等对于建筑项目能够无冲突地成功实施至关重要的一些因素。

项目的第一个阶段主要包括深度的咨询和磋商，以便项目中两个主要的参与者互相了解，并为成功且彼此信任的合作关系打下一个良好的基础。

建筑师的合同/报酬

一旦作出要实现项目的决定，那么业主和建筑师之间合作合同的书面底本便应该初拟出来。底本用来阐明建筑师工作职能的范畴及其受雇的报酬。

提示：

恰当的资金表和资金建议会反映出核心和附加的专业项目。附加的专业项目包括设计的说明、详细的记录、房屋及其功能规划的拟定等项目模式。建筑师们认为受委托的工作阶段范畴需要达成共识，同时，与他们有关的工作需求以及可能需要受委托的任何附加和特殊工作内容也最好能够取得一致的意见。

注释：

建议在每一次和业主开会之前都应该创建一份书面文件备忘录，并且分发给所有与会者。这样可以使开会之前仅仅由口头作出的决定及随后确认的意见尽可能记录下来。

提示：

项目规划过程中，建筑师面临许多责任风险。工程师和建筑师承受的责任风险不仅体现在每个实施阶段，还体现在合同的签署上。他们不但要对规划失误、投标失误和其他的合同违约行为承担责任，还不时要替承包人犯下的错误买单。目前，绝大部分国家主张职业风险索赔保险应该尽可能包含以上那些风险。

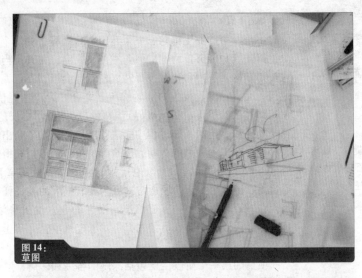

图14：
草图

图15：
工作模型

概念阶段

在项目基础、设计人员的合作关系及其主要工作范畴通过合同明确之后，建筑师就可以开始具体落实目前尚处于初步构思阶段的建筑项目。

项目文本上的关键数据这时首次转化为设计草图模式，其目的就是将实现概念构思的可能性方案转化为工程语言。>见图14

设计表现	表现设计的方式方法很大程度上留给了建筑师。建筑师通常手绘一些初期草图，或者通过简单的 CAD 图纸将他们的构思传达给业主。特别是 CAD 的表现方式，相对较容易生成许多不同视角的图纸和多种平面布局，从而给业主提供可以选择的多个方案。
初步设计	平面图、剖面图以及其他不同视角的图纸都应该按比例绘制，这个比例在初期通常设为 1：200。

建筑师在给业主展示初步设计时应该时刻牢记要采取让业主易于理解的表达方式。建筑师们在看平面图和剖面图时已有大量的实际经验，但是一些仍然不习惯看不同比例设计图纸的业主，通常仅仅把图纸看成一种抽象的表现方式，他们可能需要方向指示、出入口情况以及空间定位方面的帮助。如果建筑师希望说服业主投资他们的创意，那么他们通常需要把设计背后创造性的构思、三维的效果或地段环境的都市文脉、不同用途的房屋和空间的功能关联等情况作出更详细的说明。正如建筑师的工作就是按照业主对功能和造价的要求盖房子，业主的建议和保留意见是非常有价值的，能够刺激建筑师完善自己的工作。在项目的构思阶段，不只是业主在不熟悉的领域前行，建筑师在繁杂的项目工作期间同样会遇到一些未曾面临过的问题。>见图 15

初步设计应该满足业主的需求，传达出一种与众不同的建筑语汇或一种创造性的构思，并在和业主、最终的使用者以及其他专业设计人员的交流中进一步具体化。

成本估算	项目规模的正确阐述也要在相对较早的阶段，因为它是业主检查房屋规划需求实施方案的基础，并且业主还要在这个方案概念之上决定未来的工作。与项目相关的第一个成本估算作为项目决策的附加基础在概念构思阶段也要拟定出来，包括由项目大致规模的单位（面积、体积）确定数量单位，并乘以专项成本明细得出可参照的成本数值。成本明细即说明每个数量单位的成本是多少（如 1000 欧/m^2 毛

实例：
　　仅仅是基于房屋构造层面的建筑实践，通常不必重复说明当前的房屋建造标准、指导方针以及建筑协会明确的规定。但是对于其他建筑类型项目，如教育建筑（幼儿园、中小学、大学学院）、体育建筑（游泳池、体育馆等）、文化功能建筑（集合场所、音乐厅、露天运动场所）、办公或工业建筑群，建筑师必须要说明特殊的工程参数，例如公共建筑的特殊规范、工作流程等要求。

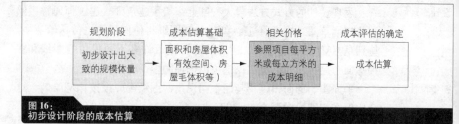

图 16：
初步设计阶段的成本估算

地面面积）。成本明细标准评估的基础是功能相似且已建成建筑所用的成本。正如项目初期会有很多无法预计到的情况出现，建筑师应该让业主清楚有关项目信息的细节只有到后期才会比较精确，初期所需成本的范畴随时都在变化。（见"概况和目标－项目成本"章节）
>见图 16

设计阶段

在业主决定实施的项目准备就绪之后，纸上构思转化为实际的工作即拉开了序幕。回溯最初的概念阶段，共同准备的成果中应该包含以下一些基础信息：

——初步概念草图（平面图、立面图、剖面图和透视图等），
——功能规划，
——空间分配，
——间数与面积分布。

基于以上这些需求，设计阶段的标准需要进一步完善，应该确定具体的结构构造，整合必要的技术设备，完工标准也应该达成共识。

注释：
因为成本的确定很大程度上取决于项目最终的规模和质量，所以项目初期的费用开销数额十分不确定。成本风险主要来自建筑项目的毛体积、毛地面面积以及有效空间方面的相关成本变量。

实例：
在一个正方形或长方形平面地段规划上，一个建筑可以有相同的体积，但是建造成本较高的立面面积与毛体积却没有直接的比例关系。正方形平面规划的建筑，其立面面积比长方形的要小得多，其成本支出也就大大地削减。在做成本明细时，类似这方面差异性的认识应该根植于脑海之中。

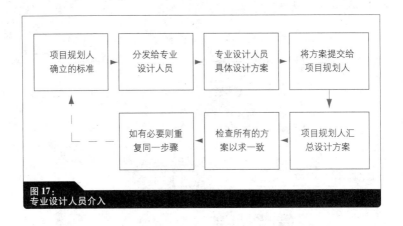

图17:
专业设计人员介入

　　一个具体的案例：体育馆在基地上的位置在"概念化过程"（conceptualization process）中已经确定。建筑的规模尺度取决于比赛场地面积和必要的辅助房屋所需的空间大小。作为整体的场馆空间与各个独立空间的关联在概念阶段已经确定。建筑师也已经将建筑外观设计的最初创意通过草图或立面效果图展示出来。

　　在这个阶段，横跨场馆的承重结构类型，应该安装什么样的供热和通风系统，在什么地方以及外墙的哪一面需要开多少孔洞已满足采光，至少还要包括什么样的材料用于建造楼地面、墙壁和天花板，诸如此类都尚未确立。

系统与整体性规划

　　因此，建筑师仍需要与业主和专业设计人员进行更广泛的讨论。"系统与整体性规划"（system and integration planning）就是适用于这个阶段的专业术语。>见图17

　　不同的体系（承重结构系统、供热系统等）须经讨论和协调；因为他们之间的交叉环节很多，所以不能孤立地考虑每个体系。例如，就场馆上空一个闭合梁的设置来说，如果没有考虑到通风、照明和其他设备所占的空间，那么从经济的角度满足场馆必须达到的高度并不是一件容易的事。如果建筑技术人员在设计最可能需要配备供热设备的时候却忽略了已设计好的朝南的大面积玻璃开窗，那么入夏的那几个月室内环境将无法忍受。因此，建筑师必须和各方面的专业设计人员一起找到安装每个体系设备最可行的解决方案，并界定每个专业方案之间的关联性，以便拿出最佳的综合性方案。

设计图纸

　　完善最初的设计需要较大比例的设计图纸。所有的平面图、立面效果图和必需的剖面图通常在下一个设计阶段都以1：100的比例绘

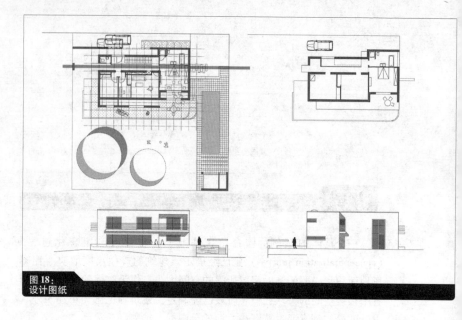

图18：
设计图纸

制，其目的不是为了详细丈量建筑所有的剖面尺寸，而仅仅是为了让建筑师脑海中建筑必要的外部尺寸、相关的房屋及其需要开孔的尺寸表现出来。> 见图18

在这个阶段，设计图并不是为了工作目标服务的，而旨在服务于业主、专业设计人员以及建筑专家。而细部设计在此时通常对工作并无意义，因为前期的规划设计在建筑师和业主、专业设计人员以及专家的讨论过程中还会有很大的变动，同时，前期考虑太多的细部将会造成工作负担过重。

平面表现

平面设计图表现的是房屋的使用功能和房屋的面积大小。根据房屋所在的位置，其周围建筑物和现存地形地貌的情况应该表达清楚。

提示：

关于设计说明和尺寸计算更具体的细节信息参见本套丛书中的《工程制图》，贝尔特·比勒费尔德、伊莎贝拉·斯奇巴著，中国建筑工业出版社2010年2月出版，征订号：18811。

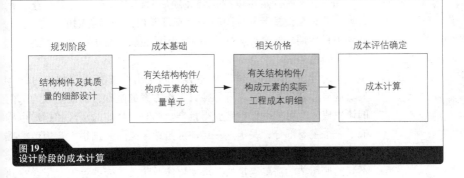

图19：
设计阶段的成本计算

概念阶段自由表达的方式在设计阶段则要让位于较大比例的统一的设计图纸，因为完整的设计图纸是构成下一阶段规划许可申请的基础，所以必须满足有关部门的要求。每个国家都有他们自己关于图纸表达方式的指导方针，但是大部分内容都是基于国家通行标准的。

图纸交换　　设计图纸必须包括很多的细部设计，以便业主了解建筑的形式风格，以及建筑师更好地与有关专家交流设计方案。因此，说明清晰不同设计师及其图纸数据转换格式（固定标准框架的DXF、DWG或PDF格式文件等）之间的界面平台是很有益处的，从而保证资料信息在交换利用的过程中尽可能高效，避免不必要的转换工作。

完成质量　　作为设计图纸流程的一部分，完成的基本质量和装订的标准必须与业主保持一致。已规划好的建筑细节描述应该由设计所有相关情况拟定出来，以作为后续工作的基础。拟订的方案将在后续设计阶段，即记录所有已决定的完成标准的阶段中不断修订。这也就意味着如果后期有设计变更，建筑师能够回溯查阅最初规定的标准，借此证明附加造价费用的必要性。

成本计算　　已定系统（承重结构和建筑科技等系统）的规划和细部设计中更详细的基础会使现有的成本评估变得更彻底。在规划的这个阶段，就交叉项目和建筑体量而言，建筑成本不需要再被评估（见"概况和目标——项目成本"章节），但是围绕单独造价的细节还需要进一步计算，这些细节包括墙体、顶棚、屋顶、供热系统和外部区域等。这也给业主提供了一个真正的基准，用以权衡建筑项目是否按计划进行，并意味着各项设计能够在各自的进行阶段作为整个规划许可的一项申请，提交到建筑权力部门。＞见图19

如果成本的确定可以细化到在随后阶段再分配给（各行业）承包人，那么在授权劳务合同时，就有可能根据项目前期的成本估算，产生项目后期成本有针对性的参照，因此也就满足了监测成本透明的一个关键性需求。

当设计做出来时，建筑师还必须制定出能够获得规划许可的计划。这些计划将是获取许可和拟订工作方案的基础，前期与权力部门的谈判成果也可能确保申请获得批准。同时，当前的法规和标准也构成了一个主要部分，关于这一点地区与地区之间差别很大。规范标准的合法有效性受到使用功能（房屋、办公和集散场所等）的影响。不同的使用功能需要满足的要求主要涉及防火、劳务保护、集合空间以及房屋构造标准等。

这些法规条款的制定是用来保护建筑的使用者、环境以及整个社会。作为设计阶段的一部分工作内容，建筑师必须建议业主知道并遵守这些规章制度。

获取许可阶段

P42 申请规划许可

有关权力部门在这一阶段决定已规划项目是否有资格获得工程许可，其关键要素在完成的比例设计中已经通过了，所需要增加的资料是尺寸链、房屋名称、面积和体积的情况。在场地总图方面，除非建筑师或建筑权力部门已经提供了相关资料，否则勘察人员就要在项目早期介入并完成场地平面图的绘制工作。建筑师要负责按照一定的秩序汇编或修订全部现有方案，为业主提供统一的规划设计标题和签名档，因为一些权力部门需要项目方案上有这种签名。同时，这也是建筑师和业主之间合同关系的一项重要组成部分。这些最初的方案还是拟定后续工作计划的基础。如果业主在项目后期要变更设计，建筑师在必要时还能够查阅到最初共同达成的设计协议和规划许可。有效的变化经常会导致额外附加费用和建筑周期的延期，而这些变化应在已经建立的规划基础中用易于理解的方式记录下来。

类型和文件

另一个与规划文件一起提交给权力部门的内容是关于业主和项目规划阶段所涉及的关键人员的个人资料，例如建筑师和结构工程师的资料。建筑项目也要有所描述，包括建筑类型（房屋、学校等）的界定、使用功能类型、建筑造价预算、房屋净面积、封闭空间和所使用的主要建筑材料。

结构工程师要记录他（她）被限定的职业资格以及项目相关的静力状态情况。根据国家规定，需要具备符合节约能源条款的证明。

根据项目的属性和规模，还需要提供隔声、防火、停车空间数量和场地使用率的证明。为了疏通上的安全，通常需要设置附加排水系统，并引导其与排水设备系统和公共市政设施相连。

建筑权力部门在此阶段依据建筑规范条款审核提交的设计方案，但并不去检查可能存在的结构上和功能上的缺陷。在建筑师正确地贯彻执行合法规范时，即使规范中存在的不合理性是由权力部门提供的，也仍然无法保护建筑师免于责任诉讼，所以给建筑师的建议就是，最好完全熟悉现行的所有相关法律和适宜当地情况的建筑规范。

建议和告知的职责

如果质疑业主的意愿不易获得许可，那么建筑师在项目初期就应该告知业主，并在必要时推荐有经验的法律顾问进入项目。在有争议的时候，建筑师不应该允许自己给出明确的法律意见，尽管他们的角色是建议并指导雇主的总负责人。

提交文件

提交完整的项目文件通常是建筑师的职责。关于必要性文件的要求在国家间和地区间均有所不同。因此，最好尽早联系建筑权力部门，以便清楚各种官方正式的规定。

一旦审核检验完成，规划许可的书面文件将会直接交与业主。作为规程，与此一起下发的还有每个管理相关规范的部门出示的细节资料。如果递交文件中有一些偏差，如疏忽了对现场经理职责的阐明，或者结构工程师职业能力的认证，那么这些细节可以后续补交。

信息和条件

规划许可的内容还包括实施阶段关于以下条目的信息和条件：
——逃生路线，
——残疾人无障碍设施，
——法定必要劳动保护，
——防火，
——树木保护，
——现场人员保护，
——其他。

> 注释：
> 如今，提交每个待批文件足够数量复印件的惯例已经确立。权威专家不可避免地要花一些时间核查这些文件，特别是文件涉及多个不同部门（消防队、交通局、贸易标准部门）时。如果提交一定数量的待批规划和申请许可文件的复印件时，能够同时分别提交给所有相关的部门，这样可以在很大程度上提高批复的效率。

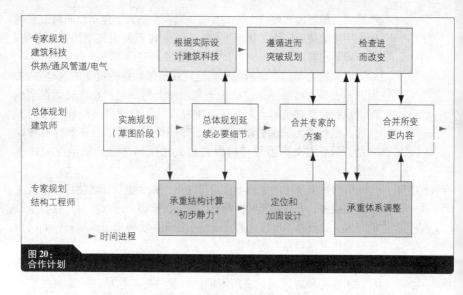

图20：
合作计划

所有这些要求在实施阶段必须得到遵守和贯彻，在建筑交付使用前，建筑权力部门必须现场检查施工项目是否满足了规划许可的要求。同时，在规划许可申请和建筑权力部门所批复的框架和内容的基础已经完成之后，权力部门还要和建筑师以及业主一起在现场检查项目的完成情况。

P45

工作计划和完工质量

前面关于项目的各个工作阶段旨在阐明和落实业主的构思。按这种思路设计好的细节方案，要被建筑权力部门检查其在许可审批进程中和合理实施过程中的实际应用性。

专业设计人员进入

实施阶段是现有方案进一步调整过程中的核心环节。相关专业设计人员也是整个过程中的关键人物。具体而言，建筑师必须告知所有的专业设计人员方案目前的状况，并把建筑权力部门的要求传达给他们。结构工程师如果在初步设计阶段没有提供建筑静力状态的初期报告，那么在这一阶段则要汇编一份方案图纸上承重结构构件主要尺寸的初步报告。建筑和电气工程师需要有关防火（火警、防火间）方面的资料，还需要掌握排水系统和相连至当地供应系统像生活用水、市政排水、煤气、电气和数据光缆等方面的需求。

规划会议

经共同认可，每隔一段时间召开一次正式的规划会议将是很有意义的。从工作的角度来看，会议安排模式可以是每周一次的固定日子，时间的确定要首先考虑到建筑师和业主的情况。建筑师负责保存会议记录和会议要点，包括作出的各项决议。会议记录可由以下内容构成：

备忘录框架

——项目标题，

——会议日期和地点，

——与会者（附一份与会者名单，每一位与会者还要登记他们的名字和联系方式，例如电话号码和电子邮箱），

——分类列表（所有出席人员和另行通知参加会议人员），

——会议记录内容。以下可按表格结构排列：

第一列：议程要点连续列举；

第二列：议程要点标题之后的内容；

第三列：职责（谁负责什么）；

第四列：完成项目该部分内容的最后期限。

最后，必须讨论记录的要点是下一次会议的日期、地点和需要出席者名单。已经确定待讨论的条目应该明确，以便参加者在下一次会议来临之前做好相关准备。

——会议记录由汇编人员签字，任何附属文件（日程表、规划文件等）都应该在列。

当然，会议记录应该如何构成有多种形式，但其目标是简洁地记录所讨论的内容，从而为以后的会议和已讨论的内容向更深层次的挖掘打下一个基础。

业主的参与

在这个阶段，业主如能参与到会议中去，并且看到所有的会议记录和设计人员之间交流函件的话，将是很有益处的。因为业主必须在细节实施方面作出大量的决策，而设计人员要为此提供必要的文件资料，还必须在成本、日程和完工质量方面给业主提供建议，以取得理想的成果。

合作

专业设计人员之间富于建设性的合作和良好的协调关系是建筑现场工作顺畅的一个重要的先决条件。在现场工作会议和规划会议进行的时候，专业设计人员合作与协调之类的会议也应该延续到施工阶段。＞见图20

> **提示：**
> 强烈建议保持规划会议和现场工作会议分开进行的方式，因为在现场工人面前，如果设计人员在之前已经认可的设计细节上再次出现争议，最后得到的结果并不会太理想。

规划次序	同时,作为日程进度表的一部分,有必要在规划设计次序上达成一致意见。结构工程师、电气工程师和建筑科技工程师需要将建筑师的设计方案作为自己设计工作的基础。他们的设计内容要融入建筑师的工作规划中。因此,彼此的工作在某一方面都是互相依存的,而且还需要一定的时间以适当的方式递交他们自己那部分的工作成果。
专业设计人员的工作	建筑师与结构工程师共同确定承重结构构件的尺寸和大样。从项目财政能力的角度来说,构造的变化是要被审核的。建筑设备工程师设计供热系统、必要的冷却和通风系统、管道工程和可能取舍的能源工程系统。电气工程师界定电的供应标准、必要的数据电缆以及照明概念。如果有其他方面的专业设计人员介入,比如声学专家、景观建筑师或室内设计师,他们会要求业主提供需求以完成所需要承担的工作。至少,建筑师自己对所有结构构件的材料、色彩和形式,还需要做出大量的决定性工作。
专业设计人员的共识	不同设计原则的交替工作和尽可能早就达成共识的必要性应该加以阐明。下面以附加楼层高度为例。
	即使在设计阶段业主已经确定了房屋的举架高度,但是作为待审批进程的一部分,楼层或建筑高度还是要依据当地情况和建筑物最大可能限高的基础上有所调整。如今,业主愿意将电缆和数据线埋入楼层管道中和插座后面,于是供热工程师的职责便是将供热系统设计安装于地板之下,这就意味着楼层的高度不太可能与最初的设计取得一致,因为地板下面设置必要的管道和供热系统所占的空间必然会明显增加地板的高度。
信息流通	这个楼层高度改变的例子表明,在已设计好的尺寸参数下,不同的需求之间未必能够互相满足。此类问题通常是由包括设计人员在内的人找到解决的办法,从而确保设计双方可以展开讨论并保障信息资料的全面流通。项目规划得越早越彻底,后期的设计和施工被扰乱的次数就越少。造价超额和施工周期延长经常都是前期规划的粗浅和不周密所致。> 见图 21
施工图	项目方案现在通常由 1:50 甚至更小的比例来详细表现。这些为所有行业准备的所谓"施工图"是在适当的比例基础上由必要的细部信息组成的。
数据交换	现在绝大部分图纸都是在 CAD 系统下绘制的,数据能够以数字模式进行交换。数据交换格式的标准目前也已经建立。项目规划的参与者们需要正确的矢量格式(DWG 或 DFX 等文件格式)以便进一步

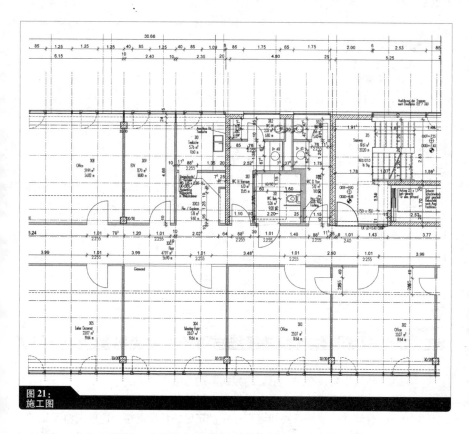

图 21：
施工图

加工数据。这些数据反映了建筑的精密尺寸，但它们的表现方式通常又有所不同。因为不同专业的设计人员使用的 CAD 系统（建筑、静力分析、建筑设备、勘察等系统）也不尽相同，因此，最好可以彼此既传输图纸打印后的像素文件（TIFF、JPG、PDF），又交换图纸的矢量文件。

图纸内容

以下所给出的项目实施的途径和方式应该被用于图纸内容的制定。清晰的绘制设计的平面、立面和所有必需的剖面是十分关键的，同时，图纸细部的逻辑表现和全部设计文件的序列编号需要在相关行业的每套设计方案中均有所体现。

P49

招标文件

投标

在获得规划许可和绘制施工图之后，建筑师必须汇编各种文件从而通过招标来实施建筑项目，之前只是体现在方案图纸中的项目信息

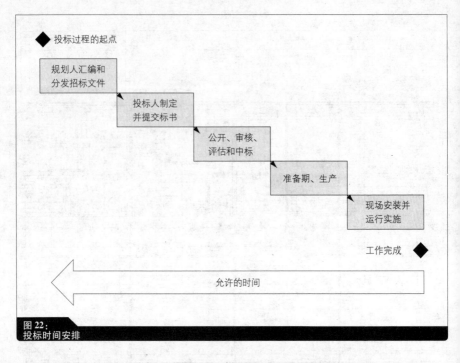

图 22:
投标时间安排

在这个阶段必须以书面的方式阐述清晰。

如果相关行业人士要想从市场的角度检验其递交的投标书内容的合理性,他们需要知道竞标项目充足的背景资料。这些资料是以招标文件的方式呈现,内容包括建筑项目总体信息、承包人条件、技术需求、特殊工作内容要求以及说明性图样。

注释:
没有必要为每一个行业提供建筑现有的整体与细部的全套图纸。一个详细的剖面图,例如屋顶阁楼或屋檐的大样图,已经可以包含建筑外立面施工人员、木工匠、屋面工和水暖工所需要的信息。其中建筑外立面施工人员负责阁楼竖柱的厚度和边饰,木工匠人需要确定檩子的定位和尺寸,屋面工确定屋顶保护层的样式,水暖工则负责遮雨板、金属板或排水系统材料的施工。

实例:
一些行业人士,如建筑正立面的施工人员,需要完整的大样图纸以满足不同详细点位的需求。这种图纸需求工作通过构架建筑清晰的结构关系,从上至下,乃至侧立面彼此关联的有机序列,为投标和连续性施工打下一个坚实的基础。

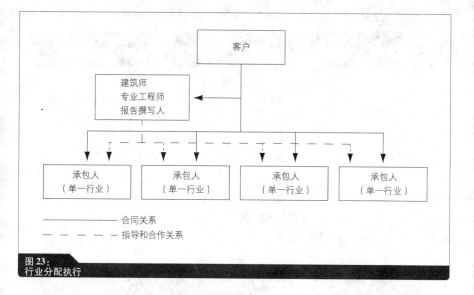

图23：
行业分配执行

质量/成本/期限

　　项目完成的质量和标准是在工作需求定义中界定的。设计人员在描述工作内容时必须把有效预算考虑进去。当授予不同的委托任务时，业主有机会先行对实际市场价格进行考察，从而与建筑师提供的项目费用互相对比。

　　在计划和拟定招标文件时，业主的时间要求很重要，分别体现在两个方面。第一个方面，建筑师必须在合适的时间制定好招标文件，并在日程表的规定时间内确定中标的承包人，以确保项目工期。（见"规划进程－投标"章节）第二个方面，作为合同的一部分，投标文件要包括承包人完成工作的最终期限。

投标进度

　　作为监测项目进程最终期限的一部分，建筑师必须首先拟定并分发招标文件给相关行业和承包人，以便他们能够在建筑周期内按照规

> **提示：**
> 关于投标和中标的细节信息参见本套丛书中的《建筑招投标》，提姆·勃兰特、塞巴斯提安·弗兰森著，中国建筑工业出版社2010年2月出版，征订号：18858。

初步措施	骨架轮廓	建筑外框正立面
现场筹备 爆破工作 土方工程 后勤工作 …	土方工程 砖石 混凝土 钢铁架 焊接 细木 脚手架 …	屋面材料 管道工程 隔热工程 抹灰工程 金属加工 玻璃装配 油饰 …
饰面修整	建筑科技	收尾工作
抹灰 磨光 总体楼面料 铺瓦 干燥施工 木工 油饰 …	供暖装置 卫生装置 通风系统 电气系统 电梯系统 媒体科技 …	建筑清洁 配锁 标牌 户外空间 建筑现场清理 …

图24：
典型的行业分工

划的时间正常开工。阐明委托工作的性质需要留有足够的时间。必须考虑到业主的要求以及其他受委托的专业设计人员们在工作上可能存在内容交叉的情况。委托有挑战性的工作可能还需得到专业公司和专业顾问的一致认可。这种工作方式获得的视角还会影响到最终的规划阶段或附加细部的设计方法。

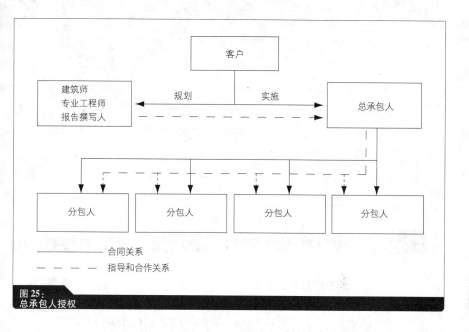

图 25：
总承包人授权

当拟定投标文件时，规划者要注意到建筑现场施工时间的顺序。对于这类建筑投标人，从节约大量必要性的组织和审核时间的角度来说，每个完整服务项目（sevice package）和招标单位可以同时进入工程。或者，作为每个招标单位的选择，工作项目可以整体委托给投标人，也可以单独给每一个独立的承包人。（见"项目参与者–承包人"章节）

如果投标与项目进程同步进行，那么工作就可以如期贯彻执行。但是工程项目内所有工作的设计和招投标工作必须在建筑开工前落实，这也就意味着建筑开工前需要给规划设计留出更多的时间。>见图 22

行业竞标
（多个行业）

项目内各行业工作范畴的界定主要由行业专家或技术公司阐明。授予项目承包权的正规方式是根据行业特点单独竞标——竞标单位需要描述合同中指定完成的工作。根据具体情况和要求，几个行业可能联合组成一个投标单位（整体中标，package award），也可能被划分为几个投标单位（行业分配，part lots）。>见图 23、图 24

行业分配投标

如果项目要求涉猎甚广（例如高速公路建设），或者有其他原因如风险分配或能力所限需要包含几家公司（几个建设阶段同时进行），那么一个行业操作可以被分解为几个行业分配。行业分配就是当一个行业操作被分解为几个部分且采用相似或相同工艺技术时的术语。>见图 25

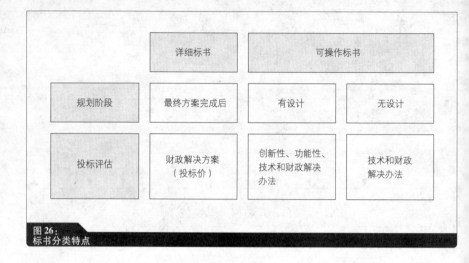

图26：
标书分类特点

总承包人授权	所有的行业类型可能由单一的承包人（总承包人）负责，其优势在于承接任务时委托合同的惟一性，以及执行项目时保有惟一的合同伙伴关系。总承包人负责协调单项工作，但作为一项规矩，每个单项的固定开销和完工日期要征得总承包人的同意。（见"项目参与者–合同框架"章节）
标书类型	建筑工程投标书的说明主要有两种操作方式。>见图26

P54

详细标书（质量清单）

每个行业的详细标书都需要完整彻底的操作方式和细节方案的说明。基于这个说明框架，每份标书根据已确定中标单位的情况分发出去，预期要达到的结果和实施过程要详细罗列清楚。承包人要在标书中获取有关质量、材料和建筑技术方面详细说明的信息。规划人必须完整且明确地描述清楚工作内容，以便每一个投标人计算投标价的前提是相同的。标书框架主要由以下几部分组成：

—标题，

—次标题，

—工作项目。>见图27

框架层：标题　　实际的投标说明被划分为几个不同的标题，分别由不同性质的工作单元组成，即各行业或房屋场地的次作业分组，这样可以使计算和审核价格变得更容易（例如：屋顶作为投标单元可以被划分为几个分项小标题，即屋顶覆层标题和屋顶排水系统标题）。

详细标书			
行业：油饰—室内			
标题：03油饰—墙体			
条目编号	短文本		
	长文本	单价	总价
	数量—单位	（up）	（gp）
03	墙体		
03.10	墙体油饰		
	外观适宜性检查， 支撑力和附加质量 清洁表面、主要的可吸收材料表面。 夹层和底层 掩饰的内层油漆：晦暗无光泽材料 颜色退晕：白色老化		
	制作：........................ （投标人资料）		
	350m²		

图27：
投标条目示例

框架层：
副标题

副标题是展开了的标题分项，特别是在大型建筑工程或建筑群项目的案例中，用副标题这种方式划分不同内容是很有意义的，并因此确定不同工作内容（例如制定建筑周期或设计结构构件）之间的差异性。

框架层：工作项目

投标说明书中最小的分项内容是工作项目，也就是描述有待完成的工作是怎样的一项内容。此项内容文本有长短之分。

长文本/短文本

短文本是工作项目标题的名字。每一个标题下面是描述实施工作的长文本，长文本的表达方式要易于理解且具备普遍性。标准的文本模式可以拿来借鉴，但也要结合项目情况进一步调整。文本中所涉及的"数量"和"数量单位"可提供样本说明，如 $25m^3$。

建筑师在每一个工作项目中都必须基于现有工作方案确立必要的数量和尺寸（m、m^2、m^3、每项25等），其目的就是提供关于工作范畴的正确信息。

P56　　　　　　可操作标书（投标纲要）

可操作标书的说明更加概括且简洁，即描述清晰整个工程项目预期达到的效果即可。原则上标书要确定一个完工日期，工作组织程序不要留给承包人，且离他们越远越好。但其实只有预期目标是必须描述说明的，一些规划工作也会移交给承包人，这在项目规划进程的初级阶段是可能被授予的。尽管如此，在具体实施项目和处理细节时，业主和建筑师在可操作标书中的影响力还是要小于详细标书中规定的角色定位，所以设计质量也就难以得到保障。如果这种标书类型被选中，那么必须在有或没有设计的标书说明之间作出选择。

没有设计的标书在构造和空间计划的基础上确定建筑的需求，其目标是当承包人拟定出投标文件时，仍然可以有设计构思的更新。如果设计已经上交，即含有设计的标书中，构造和空间计划的需求会由设计和筹备方面具体的构思补充，但是投标的可操作性被保留，也就是项目规划在技术上的执行权很大程度上留给了承包人。在可操作标书中，投标承包人承担数量风险，因为他们必须自己确定材料、建筑进程和数量，并且将这些因素考虑进投标成本中去，而详细标书的情况与之截然不同。

P56　　　　　　标书框架

原则上一个标书框架的创建方式都是相同的，无论它的形式是细节性的还是纲要性的。投标框架类似：＞见图28

　　　—正文要点；
　　　　—工程项目的总体信息，
　　　　—合同条件，
　　　　—技术要求，
　　　　—建筑现场条件信息，
　　　　—标书说明（详细标书或可操作标书），
　　　—绘图要点；
　　　　—场地平面/场地调整平面，
　　　　—平面图/立面图/剖面图，
　　　　—细部设计（如有必要），
　　　—其他要点；
　　　　—图示文件，
　　　　—样本，

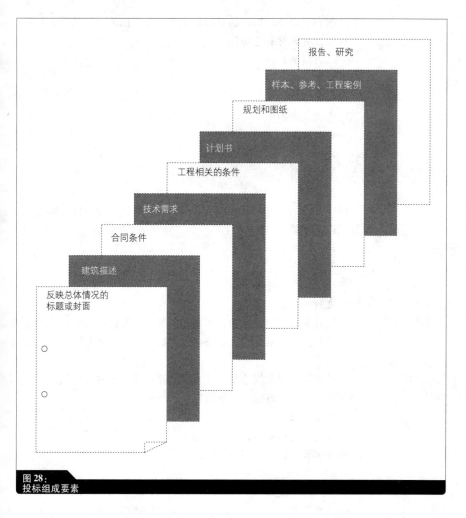

图28：
投标组成要素

——参照物/已完成的案例，

——报告/研究。

项目总体信息	封面页列举总体信息和授予形式（报告递交限期、项目完工限期、授权过程的性质等）。业主和关键个人的名字要出现在规划内容中。简而言之，工程项目的总体情况介绍要放在报告的起始页。
合同条件	一旦委托任务确定，投标人递交的文本就会变成制定合同的基础，调整后的合同样式便放入投标文件中。一般性和特殊性合同条件之间的差异要区别清楚。

一般性合同条件制定的基础是国家和国际标准，包括以下信息：
——工作的性质和范畴，
——工作报酬，
——完工限期，
——注意事项，
——责任义务，
——建筑验收，
——授权担保，
——货品计价，
——支付款。

特殊性合同条件与具体项目有关，在一般性合同条件基础之上还要补充以下信息：
——递交账单，
——支付款形式，
——次承包人劳务安排，
——支付折扣。

技术要求　　一般性和特殊性合同的技术要求，需注意要包括实践和相关工程公认的法规意义上的标准，以及在已完成工作质量之上提出的附加或更高的要求。

工程现场情况　　工程现场情况的描述是承包人在计算投标方案的场地利用计划时一项重要信息，最好包括以下内容：
——位置/地址，
——可能通达的道路，
——仓库设施，
——脚手架/起重机，
——现场供应的水力和电力，
——卫生设施。

标书　　标书是投标文件的核心（见上）。

图纸要点　　规划文件所附的图纸和草图可以帮助投标人充分理解所承担的工作。采用简化的方式递交规划文件将令人乐于接受，而在计算和核查图纸使用数量的时候采用按绘图比例分配的方式会更加简易。

其他描述要点　　随投标文件补充进工作内容描述的任何案例、影像等资料都体现在正文部分和设计方案中。

无论哪一种投标程序被业主和建筑师选择，技术人员的专业知识都是必不可少的。在这个阶段，建筑师自身所扮演的经验丰富的"建

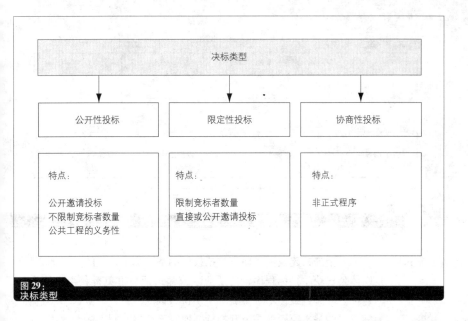

图 29：
决标类型

筑服务供应者"角色，其重要性要胜过具有创造性设计的工作角色。投标是规划和施工之间的桥梁。高质量规划的实施只有在投标阶段被全力执行的前提下才会实现。

决标程序

为了建筑工程的实施，建筑师在已经汇总必要的文件以吸引投标后，他（她）还必须找到合适的公司和技术人员，所谓合适，即他们的具体情况可以满足业主和建筑师的构思需求并予以实现。

合适的公司即那些具备必要专业知识背景，能够提供足够可靠的实力完成所承担项目的公司。就员工构成和设备仪器而言，公司在开工时拥有足够的生产力实现预算框架下的工程项目。

找到这些承包人、技术人员或建筑公司可以通过多种途径。针对小型且私人的建筑项目，建筑师可以把他（她）经常合作过的技术人员推荐给业主，这些技术人员通常都是有丰富经验的人，他们的工作能够达到预期的满意效果。

针对大型的项目，特别是公共建筑项目（国际级、地区级和当地权力部门级别的项目），建筑师有责任针对项目的不同内容，基于多种程序模式展开招投标。> 见图 29

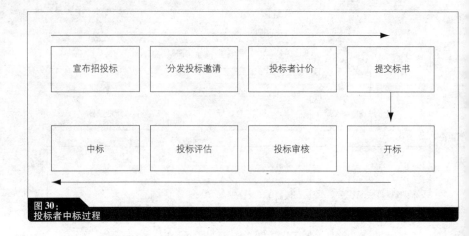

图30：
投标者中标过程

公开性投标　　　对于公共建筑项目而言，招投标的基础都是公开的。投标人经由媒体公告获悉投标程序，并从业主或建筑师那里免费获得申请文件。这种公开流程致使不限量竞争中出价最低的投标人将被选中。这种中标方式的基本原则就是公开招标，而且不严格限制投标人数量。

限定性投标　　　严格限制投标人的程序只有在理由充分且特殊情况下才可能采用。这时，有限的一些承包人直接受邀投标。在发布投标文件之前，确定所选择的公司是否原则上已经做好准备提交投标是十分重要的，否则在投标公开日期到来之际，无标提交的情况是很冒风险的。与公共性投标流程不同，限定性投标在公开之后，投标人不再被允许调查其职业能力。

　　　限定性投标的另一个与公共性投标流程不同之处是被选中的投标者们直接提交文件，无需任何费用。不过在每一个项目案例中，都需要公开标价。

协商性投标　　　对于公共建筑项目来说，协商或私人委托的投标程序必须详细阐明情况。这种投标竞争，即使在业主只与一个投标人协商的情况下，其要求也是很严格的。＞见图30

> 提示：
> 　　当然，私人业主没有义务必须选择以上所说的某种投标程序，在委托建筑承包合同时，倾向于保持一种开放的或可协商的程序。但是仍然强烈建议要求多一些投标人参与，目的在于可以获得与市场价相符的投标价格，因为投标价的出入可能会非常大。一个投标人如果意识到其他人也在准备投标时，往往会提交一份更加合理的报价。

油饰招投标					
限定性投标					
价格比较明细表					
明细条目	内容	数量/单位	投标人A	投标人B	投标人C
1.10	清洁	50m²	€ 3.50 € 175.00	€ 4.00 € 200.00	€ 4.20 € 225.00
1.20	初步油饰	50m²	€ 2.50 € 125.00	€ 2.40 € 120.00	€ 2.90 € 145.00
1.30	内层油漆	100m²	€ 4.50 € 450.00	€ 5.00 € 500.00	€ 6.20 € 620.00
1.40	夹层油饰	100m²	€ 0.50 € 50.00	€ 0.80 € 80.00	€ 1.20 € 120.00
合计	标题1 标题2 标题3	净利 净利 净利	€ 850.50 € 1320.00 € 720.00	€ 915.00 € 1280.50 € 835.00	€ 1280.20 € 1450.90 € 830.50
合计		净利 毛利 折扣 扣除 总数合计	€ 4850.50 € 5771.50 — — € 5771.50	€ 5210.50 € 6200.50 — 5% € 5890.48	€ 6160.30 € 7330.76 — — € 7330.76

图31：
价格比较

公开日期/提交

公共决标程序中会说明统一提交投标的日期，届时，所有投标必须完成提交。在指定的时间，在所有投标者面前公开中标方案。投标提交延迟者不予考虑，因为机会的透明性和平等性是最重要的。

核算和技术审查

一旦中标方案公开，建筑师必须审核和评估方案的细节。正式检查与中标方案的签约是否正确，以及中标方案中有无删改和增加的内容。中标方案被拿来与其他方案互相比较的审核工作是非常必要的，否则与同样满足要求的竞争者相比，中标者会获取不公平的利益。审查核算工作不仅仅关注投标总价的比较，关键是比较每个单价和计算方式，因为单价会成为签署合同的基础数据。

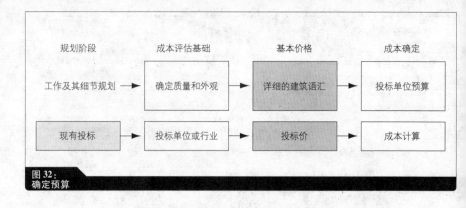

图32：
确定预算

价格比较　　　　　　拟定价格比较列表的目的是方便展示投标报价情况。>见图31 所有的投标价格都被输入计算机程序中进行比较。原则上，这项工作可以使用电子制表软件如 Excel 的程序设计，但是中标货品计价（tendering – award – invoicing）程序已经普遍通行，它便于汇编价格比较列表。价格的最大值和最小值可以清晰地标注出来，使用百分比排除差异，并能够识别投标单价之间的偏差。同时，还可以首次显示投标人计算的类型以及投标文件中可能存在的文本误解。

授权建议　　　　　　完成审核之后，建筑师提交一份委托提案给业主，提案需要使委托项目能在财政预算内分配下去的期待变得可能。>见图32

成本控制　　　　　　一旦所有来自各专业或单位的投标中标，便有可能比较已计算的和已批准的预算。数据比较的结果非常可靠，能够帮助业主判断建筑师之前计算过的成本是否准确。当成本计算仍处于规划阶段时，那么基于投标单位的价格汇总预算将是至关重要的，因为这样得出的成本才易于理解。如果招投标继续进行的基础是独立行业或行业组群（trade lots），便有可能对随后投标中造价超标的情况加以干预和控制。

> 提示：
> 　　单价是投标人对于工作项目每一个数量单位价格的说明，构成了项目后期所投入实际工作报酬清单的基础。总价是将已标明的数量或实际列举的数量的单价累加而得。投标实价就是所有总价之和。

授权会议	在业主将项目委托给承包人之前，最好能召开一次委托授权会议。因为在公共建筑项目中，投标价格的后期协商是禁止的，但是对于一名私人业主来说，却是获取实惠价格的有利环节。所以，作为公共投标程序的一部分，授权会议有助于澄清关于项目施工生效问题和可能存在的价格选择问题。
工程合同	处理业主和承包人或技术人员之间实际工程合同的诸多事务并不是建筑师的主要职责所在，尽管建筑师的确需要给业主关于合同条款方面的事宜提供建议，例如最后期限、合同处罚、报酬形式、折扣、授权期、安全保障措施或类似的一些事务。
时间要求	以上关于时间的参考条款需要准确起草进投标文件（见"规划进程－投标"章节）。投标程序的日程表也必须规划出足够的时间。

就公共建筑项目而言，法定的最后期限作为委托进程的一部分必须得到确认。在所委托项目的投标明细单已经发表和承包人已经申请投标文件之后，投标人需要足够的时间汇编他们的报价。在以上这些文件返还之后，规划者也需要一定的时间评估这些报价，以便更尽职地审查。如果公共合同的委托程序是由行政委员会管理，那么还应该考虑到会议时间和起草这些会议所需文件的最后期限。因此，根据业主类型的不同，起草投标明细单到委任承包人之间的时间周期能延伸到几个月。一些行业需要在建筑现场工作前准备绘图工作室，预订和处理的建筑材料从订货到交货会需要较长的一段时间，这种情况也应该考虑进日程安排中去。

P65	**施工**
	目前为止，所有规划行为的描述最终要达到的实际目标即：工地现场实施规划。
工程开工	公共项目中，建筑实际开工日期是业主非常好的一个庆祝理由，因为从这时起，社会大众和现场邻区真正意识到建筑项目正式的启动了。私人业主更乐意在工匠安装屋顶三角桁架后举办一个"平顶宴

> 提示：
> 关于建筑项目施工的更多细节参见本套丛书中的《场地管理》，拉斯－菲利普·鲁施著，中国建筑工业出版社预计于 2010 年 10 月出版。

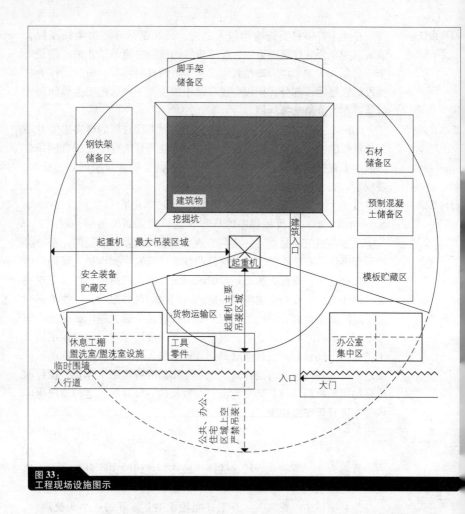

图 33：
工程现场设施图示

会"（topping – out party），然后就是接下来的建筑开放。如果是一个重要项目，公共业主为告知普通大众，经常通过第一次挖掘地面以及安排"挖掘机开挖"（digger bite）仪式来启动项目；或者如果该项目足够大，还会安置一处奠基石以示开工。

工程现场设施图示

项目启动后的第一步措施是安置和保护好建筑现场。一定比例的现场平面图应该绘制出来，用于标识建筑占地面积的尺寸、场地围墙的位置、入口大门、货运区、储存设施、工作区、脚手架区、起重机的位置及其悬臂的回旋半径，以及休息区、垃圾堆放区和挖掘坑的位置。建筑能源和供水系统的连接点必须确立并在合适的位置上绘制

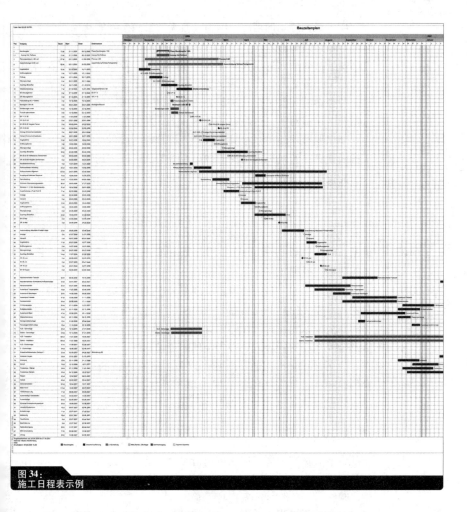

图34：
施工日程表示例

清楚。一处建筑工地在施工期间会对周围环境造成极大的影响，所以必须与当地权力部门达成施工协议，并应该与周邻社区打好招呼。当建筑实际开工时，应该在合适的时间通知权力部门。> 见图33

日程表 　　建筑现场经理最重要的工作内容之一是起草一个日程表，通常是以柱状图表的方式表示（见"概况和目标－项目期限"章节）。这个图表被用来协调主要的建筑进程和行业间的配合，便于记住各自阶段的完工日期。同时，很重要的是体现出各个行业的工作被整合在一起，并有效地互相配合。> 见图34

　　如果在项目完成后再回顾工作日程表的第一个版本，很容易发现有背离了最初规划的情况。因为会有无数意想不到的状况影响着日程安排，包括糟糕的天气、授权问题、材料运输堵塞，甚至施工过程中可能出现的破产状况以及承包人导致的误工等等因素。日程表的制定基础是工作的正常顺序，某一个环节出现误工都会导致随后施工的延迟，乃至影响到整个建筑进程。现场经理必须密切关注所有的环节，以便及时采取对策，并且提供延迟完工日期的信息。

日程干扰因素

如果要合理维持一个日程表，应该考虑到项目进程中所有必要的干扰因素。在建筑周期结束前如果有短期延迟施工的情况，那么就不大可能按照业主规定的日期完工，所以必要时应该尽早采取干预措施。

干预措施的形式可以是提高工人的工作能力，也可以是延长工作时间。将工作划分成几份也不无裨益，这样后面安排的工作就不会被阻断。对于部分工作，还可以在后期继续完成，这样至少不会影响到过渡限期或后期工作。

现场总体指挥

项目施工中是否需要现场经理很大程度上取决于项目的规模和复杂性。不需要独立的现场办公室，建筑师通常可以自己负责的项目是那种小型的家庭工程，而那些大型工程项目则需要建立现场办公室，以及全程在场的一个或更多个经理的现场指挥。

为了与承包人沟通，现场经理需要所有最新的设计图纸和建筑合同，包括已经委托工作的明细单。通过图纸细节熟悉承包人，对于现场经理来说是绝对有必要的工作，如此方可根据要求直接指导工作。

现场会

大型项目的现场办公室还被用于定期召开现场会的场所。现场工作的专业人员和公司都应该被邀请参会，这样做既必要又符合工作进展的要求。会议内容应该在日常会议备忘录标准的基础上做记录。
(见"规划进程—工作计划和完工质量"章节)

对于施工进程中所有新介入的参与者而言，有必要在一开始就交换联系方式，然而也要注意，这些由现场经理负责的信息其流动性是很频繁的。

施工日志

现场经理负责记录施工日志。施工日志是作为现场已完成工作的记录，经常在后期被用来提供证据。下面是日志中非常重要的条目：
—日期，
—天气状况（气温、白天的时间等），
—现场工作的公司，
—每个公司的工人人数，

——已完成工作的性质，

——定购、说明和认可情况，

——设计移交、样本等，

——材料交付情况，

——特别事件（参观、延工、事故等）。

在现场

正如已经建议过的情况，现场经理是否在场很大程度上取决于手头项目的进展。举例来说，只是记录施工日志并不需要现场经理每天都来工地。如果现场经理需要或必须出现在工地现场，只不过在日志中提及日期即可。现场指挥的同时，还必须遵循委托许可、工作方案和工作明细单的内容，遵守实践认可的规律。

毫无疑问，现场经理在施工的重要和关键阶段必须出现在现场，例如用恰当方法锚固承重结构以及挖掘工程时，要监管正在施工的密封和隔离工程、加强材料的采用、按质量标准要求浇筑混凝土等情况。如果隔热设施和结构的隔声设施正在安装，也需要现场经理现场监督。

对于简单和例行的工作，现场经理不需要在场。有经验的经理能够合适地掌握自己在工地的时间。

健康和安全官员

现场经理有责任注意建筑工地的交通安全和灾害预防。在这方面，他（她）得到健康和安全协调员的帮助，协调员定期巡视遵循规则规定的事故预防工作，如果没有事故发生，要给现场经理指出潜在的危险。（见"项目参与者—专家组"章节）

专家现场管理

现场经理的职能范围有限，本质上就是协调关系。就工地现场工作的专业公司来说，主要是由专业工程师指导，而不是现场经理。同时，现场经理也绝不会因为行业操作超过他（她）能力所及而雇用专家管理。

提示：

即使是建筑师，如果他（她）与现场经理不是一个或相同的人，就不应该直接指导工人，但不包括现场经理本人。建筑施工进程非常复杂，以至于每个相关个体都不能对所有互相关联的程序和事情有一个完整的掌握。为了避免不同组织单位之间的矛盾，所有信息必须经现场经理传达，现场经理则负责协调所有的关系。

实例：

如果现场经理认真负责，在一处著名的建筑项目工地一周出现两次，并且能在遵循期限的同时保障工程质量，那么业主就没有理由反对他（她）出现在现场的时间。但是即使现场经理一直在工地，协调失误也会悄然发生，或者还有一些事情有待改进。管理现场经理的建筑师也会将"成功"归于现场经理在管理工地期间的作为。

协调	现场经理是协调具体工作和行业专家之间关系的关键枢纽。两个例子说明"现场经理"作为控制中心角色的重要性。第一个案例是只有具体负责的现场经理能够协调安置清水隔断墙和墙两侧板材之间,安装电气设备或附加管道工程与随后的墙体封闭和抹饰之间,以及电气开关最后装配之间的交叉配合问题,因为各行业自身没有理由考虑其他后期上马的行业做的工作。另一个需要更多协调工作的典型案例是安装于地板下的供热系统和找平层下电缆管道的平行设计。所有相关行业必须知道后续工作的时间安排,允许留出找平层干燥的时间和制定供热系统安装草案的时间。既实际又合理的时间安排次序方案必须经由具体负责的现场经理和专业设计人员共同制定。
验收	项目实施过程中,建筑师和雇用公司所完成的工作都要给出评价。验收工作之前,业主若是一直说明他(她)认可工作完成的情况,那么这个项目实施过程便可以有相当多的合法成果。建筑的验收通常是在外观施工完成或最终收尾阶段,在这个时候,项目对批准文件或当前建筑法规的遵循与否尚需检验。官方供应网络连接的验收则是由负责水、排水系统、电力、煤气等供应部门具体执行。供热系统的验收是由当地优秀的烟囱清扫工承担。特殊的技术设施,例如运输系统或电梯需要由技术监控设备独立验收。
执行索赔和保修期	以前,业主针对承包人或行业人士的工作有一条"执行索赔"(compliance claim)的要求,其中损失风险一般由承包人承担。"损失风险"(Risk of loss)的意思是指项目在验收之前发生意外或损害事件,进而所做的无报酬的弥补工作,这种工作所承担的风险即为损失风险。
评估	规划工作的评估也不可或缺,其评判依据是不同地区自己的法规规范和建筑验收方式。同时,在建筑完工之后和移交使用之前,要评估所完成的工作是否遵循了最初审核通过的设计方案。
	保修期开始于验收之时,这方面内容详见下一节。在保修期或有效期内,发生任何的缺陷错误都必须得到纠正,只要这种纠正方式及其开销不要陷于失衡的状况即可。一旦缺陷得以很好的补救,保修期的起始日期就再一次启动。
调查	各行业人士通常在工作任务结束和工作验收前就开账单给业主。因此,通常会采取"分期支付报酬"或"部分支付报酬"的方法,其支付依据是建造的进程和已完成的工作内容。依据的基础则是最初的工作计划。如果计划中发生特殊情况又没有记录,或者是有违背工作计划的情况,那么需要在现场和工人一起调查发生变化的地方,以

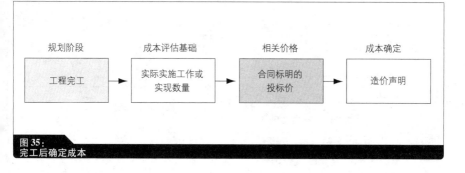

图35：
完工后确定成本

审核账单　　便能够正确地核对所提交的账单。
　　　　承包人在提交正式账单时，形式上是交给业主，实际上是交给建筑师。建筑师负责在职业和计算方面审核账单，他（她）改正账单的错误后，将账单和修改说明一起交给业主。在核对时，建筑师必须一直牢记自己是业主的代表，向着承包人核对不是他（她）的职责，替业主着想才是他（她）的责任。最终，账单修改说明只不过是业主应该支付报酬的一个参考。

成本说明　　随着账单现在逐渐从各行业独立出来，业主有必要为此准备好大量的可用资金，以及为规划项目所需支付的酬金。业主因此逐渐将兴趣放在对当前成本费用情况的关注。项目的预算评估通称是有限的，其金额取决于信用存款、补助金或建筑委员会的许可，因此预算不能随意超支。将作为建筑合同提供的委托金与实际账目总额即各行业完工后的账单总开支进行比照，对业主而言十分重要。进一步说，如果各项费用支出保持监管，将有可能有效地干预成本实际超支情况。因此，随着建筑施工进展，最初预算成本将会与完工后的实际成本非常接近。整体施工完毕后，最终的造价总结将出现在成本说明中。＞见图35

移交　　　　工程验收后，如上所述，各方面专家和行业人士完成的项目最终移交给业主。这个过程虽说不必伴有建筑审查巡视，但是业主确实有权要求所有委托的设计文件均要移交。这份文件包括一套完整的工作计划、设备安装图、技术操作资料、修补加固方案、验收协议和证明。文件量的多少取决于建筑规模，但是文件本身的汇编方式必须以能够正确使用为根本。所有有关后续维护工作、使用后的重新调整或改变等方面的信息资料必须有效。同时，专业设计人员还需要提供有效的必要性资料给业主。

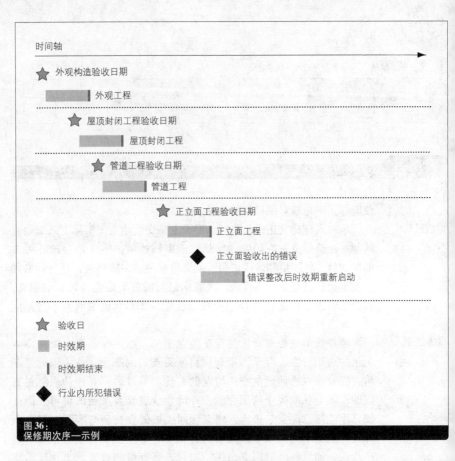

图36：
保修期次序—示例

开放

建筑师的职责绝不是在项目移交后就可以卸下的，仍可能有一些工作有待完善，承包人已完成的工作中业主不满意的缺陷必须被监管和处理。

通常，建筑最终的完工应该值得适当庆祝，通过展示具有象征性意义的钥匙，建筑师将建筑正式移交给业主。在公共项目的案例中，这种场合可以吸引大众的关注，并提升项目的受重视度。

除此之外，施工进程中每一个相关人员承担的义务和给予的服务都应该得到认可。建筑师也应该赞颂业主，至少应该将自己拿到的酬金归功于业主良好的资金筹备状况，无论是私人业主还是公共业主。

保修期

建筑师在建筑完工之后仍能继续为业主提供有价值的服务。建筑师将独立于以上描述的移交文件之外的一份详细且综合性的文件，包括结构记录和规划过程中所有使用过的数据分析的文件交给业主。这么做并不意味着起草新的方案，提出新的规划工作，如建筑展示计划，而仅仅是将项目规划进程中所有处理过的文件整合在一起。

建筑移交之后所需工作的一个关键部分是监督每个行业完成工作项目的保修期。所承包的项目几乎都是在不同阶段完成，所以也就在不同的阶段交工，保修期即从交工之日算起。不同阶段交工的时间根据是合同的规定，因此仔细监督所有项目的保修期是至关重要的。建筑师必须在保修期开始之前全面快速地检查一遍，查看建筑的每一处缺陷以及所有的结构构件。这个检查过程并不必采用特殊的检查方法或设备。>见图36

如果检查揭露了问题，则必须通知业主。允许选择一个合适的时间纠正这些问题。

建筑师还要对建筑项目造价的指导方针做最后的检查，涉及平方米、立方米等数量单位或每个行业情况等诸方面，这样做对建筑师自己也是很有益处的。检查所完成项目负担的成本费用和实践完成工程所需的时间，对于建筑师规划未来的项目是很有用的参照。

保修期需要做的工作会花费大量的时间，这与所付的费用相比不成比例。但是这方面时间和精力的耗费可以有效地控制，至少可以依靠出色的规划、正确选择有能力的公司和工人，以及认真尽责的现场经理等多方面的帮助来实现。

结语

当建筑师开始做工程项目规划的时候，他对能否成功实现这个项目便负有责任。项目经理的职责是必须倾力关注于项目每一个阶段的最初目标。规划的需求，就成本、日程表和质量标准而言，都是同等重要的，也都是工作必须经常面对的。

建筑师是业主在项目的所有阶段都需要联络的专业人士。他（她）为业主提供解决问题的建议，包括建筑规划阶段的问题，以及有关技术、经济、创新、市政和生态等方面的问题。作为具备资格认可的协调人，建筑师的职责任务是协调建筑项目中所有相关人员诸如专业设计人员、承包人、工人、当地权力部门和人士之间的关系。

建筑师按照逻辑顺序诠释规划的每个步骤，在工程项目规划的每一个阶段和业主共同作出相应的决定。

建筑师在项目实施阶段检查工作开展的次序，以确保规划阶段有序的设计目标全部得以实现。

每一个领域都会有特殊情况的增加，所以实现一个工程项目需要规划者具备广博的知识和全面的能力。作为一名建筑师是很有吸引力的，不仅因为这个职业可以创造出多种设计可能，很大程度上还因为这个职业充满了挑战。正如之前章节介绍的一样，经济实力和策略技巧尤为重要。另外，建筑师在他们工作期间还必须解释所有相关的法定事宜。团队工作的能力、与不同类型人打交道的能力都是建筑师需要具备的高标准的社会能力。建筑师还会与项目过程中大量的职业伙伴互相接触，无论是项目相关的合伙人还是从事具体实操工作的专业人士。每一个建筑项目都意味着要与全新且有趣的工作领域中潜在的规律相融合。同时，规划者至少要具备一个显著的能力来说明他（她）的设计和构思是对社会负责的，并正在促成设计。

实现一个工程项目，从初始构思到完成移交，几乎需要涵盖一个建筑师能力范畴的所有方面。每一个新的规划阶段都会带来新的挑战。规划者的成就在一个项目的成功实现和有效使用中得到了肯定和彰显。

附录

图片提供者

图 10、17、20、29	贝尔特·比勒费尔德，托马斯·福伊尔阿本德
图 22、24、26、28	提姆·勃兰特，塞巴斯提安·TH·弗兰森
图 23、25、30	乌多·布勒肯，贝尔特·比勒费尔德
图 33（工程现场设施图示）	拉斯—菲利普·鲁施
其他图片	作者本人

作者

哈特穆特·克莱因（Hartmut Klein），德国布莱施高县内弗赖堡市的硕士工程师、开业建筑师，擅长管理公共建筑工程项目。